Particle Accelerators: From Big Bang Physics to Hadron Therapy

Ugo Amaldi

Particle Accelerators: From Big Bang Physics to Hadron Therapy

With the Collaboration of Adele La Rana

Translated by Geoffrey Hall

 Springer

Ugo Amaldi
CERN
Europ. Organization for Nuclear Research
Geneva 23
Switzerland

The translation of this work has been funded by SEPS SEGRETARIATO EUROPEO PER
LE PUBBLICAZIONI SCIENTIFICHE, Via Val d'Aposa 7, 40123 Bologna, Italy; e-mail:
seps@seps.it; www.seps.it.
Translation from the Italian language edition of: Sempre più veloci, Copyright © 2012
Zanichelli editore S.p.A., Bologna [6331].
Authorized translation from Italian language edition published by Zanichelli. All Rights
Reserved.

ISBN 978-3-319-08869-3   ISBN 978-3-319-08870-9 (eBook)
DOI 10.1007/978-3-319-08870-9
Springer Cham Heidelberg New York Dordrecht London

Library of Congress Control Number: 2014955693

Cover figure: Three images showing the three aspects of particle accelerators discussed in the book:
machines, particle physics and medical applications.

Credits: 1) Digital painting, "In search of the Higgs boson: H -> ZZ": Xavier Cortada (with the participation
of physicist Pete Markowitz), digital art, 2013; 2) Artistic view of LHC (Geneva): © CERN; 3) Proton
treatment room at the Massachusetts General Hospital Francis H. Burr Proton Beam Treatment Center
(Boston, USA): Courtesy IBA

Printed on acid-free paper

Springer is part of Springer Science+Business Media (www.springer.com)

To Clelia

Preface

The idea for this book grew gradually over time, taking shape from a long-standing enthusiasm for particle accelerators which originated when I took my first steps in the world of research. How to approach this theme became clear to me when, many years ago, I read what had been written by "Viki" Weisskopf, Director General of CERN from 1961 to 1965: *"There are three kinds of physicists, namely the machine builders, the experimental physicists, and the theoretical physicists. The machine builders are the most important ones, because if they were not there, we would not get into this small-scale region of space. If we compare this with the discovery of America, the machine builders correspond to captains and ship builders who really developed the techniques at that time. The experimentalists were those fellows on the ships who sailed to the other side of the world and then landed on the new islands and wrote down what they saw. The theoretical physicists are those who stayed behind in Madrid and told Columbus that he was going to land in India."* The scientific adventures and discoveries I recount in this book have thus as main characters the machine builders, neglected heroes of the history of physics who rarely appear in the many excellent books on particle physics and cosmology published so far.

During my career, I carried out experiments in fundamental physics at six different particle accelerators (two in Italy and four at CERN), and I have also published many papers dedicated to this type of instruments and have taught the subject in university courses as well. I have had the great good fortune to meet, and in some cases to work side by side with, several among those persons who laid the foundation of these extraordinary machines, which are capable of breaking down matter into its fundamental constituents but also serve as an invaluable means of diagnosis and treatment of cancer. This double aspect of instruments, at the service of knowledge on the one hand and in the service of health on the other, renders them a perfect example of how pure science and applied science are intimately interrelated and indivisible.

The first edition of this book was published as a pocket-sized paperback in Italy in October 2012. The following year, it took second place in the National Prize for

Scientific Communication awarded by the Italian Book Association, an event which encouraged me to expand the content for a successor edition, including material and personalities which I had to regretfully omit from the first. The result is that the English version has grown to become a volume twice the size of the Italian one, with many more figures and citations.

For both editions, the collaboration of Adele La Rana, physicist and scientific writer, has been invaluable; she provided attentive and critical rereading and innumerable suggestions for much improved explanatory clarity and took responsibility for the search for artwork and the various editorial phases.

The English edition has been much enhanced by the sensitive translation of Geoffrey Hall, who not only rendered with clarity the spirit of the text but also contributed – with helpful criticisms and suggestions – to correct errors and inaccuracies and to make the text more comprehensible.

My gratitude is also extended to Alvaro De Rujula, Frank Close and Gino Segrè, who as scientists and communicators have provided me with important external perspectives; to Marco Durante, Michael Goitein and Gerhard Kraft, for their accurate and valuable criticisms of the chapters dedicated to accelerators for medical applications; to Guido Altarelli, Giorgio Brianti, Roberto Orecchia and Sandro Rossi, for the careful readings and suggestions; and to Federico Tibone, former editor of the series in which the Italian edition was included, who contributed with corrections and advice to the later improvement of the English version. I am particularly thankful to Ramon Khanna, my editor at Springer, who believed in the project from the outset and who provided many suggestions and has been patient with me during my many delays.

Finally, I would like to thank Paolo Magagnin, for having helped me produce several complex, previously unpublished figures included in this book, and Daniele Bergesio, who lent his enthusiasm for photography to depict the CERN locations cited at the start of each chapter.

Geneva, Switzerland

Ugo Amaldi

Prologue

The athletes who participate in the Olympic Games run and swim ever faster, even if, with the passing of time, it becomes ever more challenging to beat former records. The same can be said of the particles that travel at (almost!) the speed of light in accelerators, those microscopes of the infinitely small. Everyone understands the motivations of athletes and Olympic organisers; physicists' reasons are instead much less clear to non-experts.

The question "why accelerate particles?" has become even more frequent since – in July 2012 – physicists from the *Large Hadron Collider* (or LHC) at the CERN laboratory near Geneva announced the discovery of the *Higgs boson*. It is a question which I will try to answer in this book, beginning with the history of a few important physical discoveries and some of the main characters involved.

As we will see, it is a story in which research on fundamental aspects of nature – from the structure of matter to the origin of the universe – is interwoven with applications of great practical value, in particular for diagnosis and treatment of our illnesses.

Subatomic Microscopes, Particle Factories

The important development of particle accelerators, devices that usually have a circular shape, began about 80 years ago.

Initially, for some decades, they were used to study the structure of matter: fast particles, once accelerated, were directed onto a target, for example a small piece of metal. Observing the products of the collisions provided information on the structure of atomic nuclei in the bombarded material. It was rather like exploring the contents of a darkened room by throwing a lot of marbles into it and observing the rebounds.

Subsequently, the attention of physicists turned to the new particles produced in the collision between a fast particle and an atomic nucleus. Energy can be transformed into mass, as predicted by the relationship $E = mc^2$ discovered by

Einstein, and the energy released in collisions frequently gave rise to the creation of new particles. Most of these particles are 'unstable' in the sense that they live very briefly and then give rise to two, three, four... particles of smaller masses. These unstable particles, which before 'decaying' into other particles survive for less than a millionth of a second after the collision, are not found in the matter making up the world around us and can be studied in detail only by producing them artificially with an accelerator.

Thus accelerators, as well as being 'microscopes' of the subatomic world (that is of the nucleus and whatever exists inside the nucleus), can also be viewed as 'factories' of these particles that – having mass, and therefore energy, greater than the particles which make up ordinary matter – are unstable and decay extremely rapidly, transforming into the well known stable particles.

However, there is a *third* reason for studying the fleeting existence of particles that are so difficult to produce and observe; physicists would like to interpret the picture of Fig. 1. This is not just any picture: rather it can be described as the most ancient 'photograph' that we will ever have of our universe. Let us understand better what that means.

Why We Cannot See the Big Bang

When we point the most powerful telescopes deep into space, we observe light originating from stars of our galaxy and from other galaxies. With the naked eye we can see only some thousands of stars in our galaxy, that in reality contains several hundred billions of them and, even though the other galaxies of the universe are also numbered in hundreds of billions, we can hardly distinguish Andromeda, the closest one.

The light emitted by the farthest galaxies has travelled for billions of years before being detected by telescopes, so the photographs describe today how those galaxies were billions of years ago; in the meantime they could even have disappeared. Hence the furthest galaxies are the oldest and those closest, the youngest; comparing images, therefore, it is possible to study the evolution of the galaxies with time.

But there is more; in the last century it was discovered that galaxies are moving away from each other, and this observation shows that space itself is expanding; it is as if galaxies were tiny specks of paper stuck to the surface of a balloon that is being inflated. Extrapolating this motion in reverse, if the balloon were deflated, one can conclude that about 14 billion years ago all the energy of the universe occupied a tiny volume and the temperature of the matter contained within it was extremely high; this was the universe produced in the Big Bang.

A millionth of a microsecond after the Big Bang matter had a temperature of millions of billions of degrees. Then however, with the passage of time and the expansion of space, the temperature gradually decreased. There are places in the universe in which the temperatures today are still high; for example the photosphere

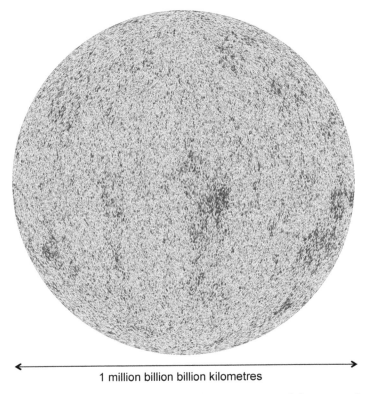

1 million billion billion kilometres

Fig. 1 This 'photograph' shows the distribution of the temperature of the gas constituting our universe when it was 380,000 years old. It was obtained in 2013 by the Planck space probe which had been sent into space 4 years before by the European Space Agency (ESA). The different colours represent tiny fluctuations in the temperature of the gas, mainly hydrogen (Courtesy of European Space Agency, Planck Collaboration)

of the Sun is around 6,000 degrees, and its interior reaches 10 million degrees. However, if one could carry a thermometer into the space between the galaxies – where the heating due to the light emitted by stars and intergalactic matter is negligible – the measured value would be −270 °C, or 3 K.[1]

Up to temperatures of several thousand Kelvin, matter is solid or liquid or gaseous, as we know from experience on Earth. Heating beyond that begins to produce a fourth state of matter, the so-called *atomic plasma*, in which continuous and violent collisions between atoms strip some of the electrons from the nuclei around which they normally orbit.

Increasing the temperature actually means to increase proportionately the energy with which atoms of matter collide, in that frenetic and uncoordinated dance that physicists call *thermal motion*. At 10,000 degrees, the collisions detach electrons

---

[1] Physicists prefer to use as a reference the minimum possible temperature, zero degrees Kelvin or 0 K, equal to −273 °C.

only from the lightest atoms, such as hydrogen or helium; to strip electrons that
orbit close to the nuclei of heavy atoms like iron it is necessary to exceed millions of
degrees.

In any case above 10,000 degrees, a significant fraction of the plasma is made up
of electrically charged particles, in particular free negative electrons and residual
positively charged atoms deprived of one or more electrons, and called *ions*. These
electrical charges immediately absorb the packets of luminous energy, or *photons*,
which are continuously emitted by atoms and ions themselves at these high
temperatures; an atomic plasma is therefore not transparent to light, like a slab of
iron.[2]

Thus, the *primordial cosmic soup* of particles – which shortly after the Big Bang
was at billions of billions of billions of degrees – cooled while expanding, but
remained impenetrable to the light emitted by the same particles until the temperature fell below a few thousand degrees.

Theories of the birth of the universe are called *cosmological models* and today
are supported by numerous precise measurements. The models tell us that starting
only 380,000 years after the Big Bang – when the cosmos was about a thousand
times smaller than at present – the temperature became low enough that most
electrons became bound to atoms that are electrically neutral and the whole
universe became transparent in a very short time. Matter consisted mainly of
hydrogen atoms, hydrogen being the simplest atom, made of a positively charged
proton encircled by a negative electron.

Thus, when the temperature decreased to the present level of the Sun's photosphere, the cosmos became the source of an extremely intense white light, whose
unabsorbed photons began to travel in every direction at the speed of light. Instead
the light emitted *before* this instant, absorbed by the primordial cosmic soup, will
never reach our instruments, however sensitive they are; looking from Earth we are
not able to 'see' the Big Bang.

Relic Radiation

To understand the significance of Fig. 1 ignore, for the moment, the expansion of
the universe and imagine that, for a brief instant, a dazzling white light were to be
emitted from every point of an enormous sphere of extremely hot gas, at whose
centre the Planck space probe – taking the picture – is found.

Later – say a thousand years after the instant of emission – the photons which
have travelled undisturbed for all that time arrive at its detectors, having been
emitted by atoms which were located, at the moment of emission, at a distance of

---

[2] Iron is opaque to light because two of 26 electrons of each Fe atom move freely within the metal,
and therefore 'devour' every photon that penetrates. Instead glass is transparent because all the
electrons are attached to their atoms.

1,000 light years away. The images then captured by Planck therefore describe the state of a thin layer of gas – of 1,000 light years in radius – as it was 1,000 years earlier.

After a million years the photons which have travelled a thousand times further arrive, corresponding in the images to gas which was located one million light years away. After a billion years the same instruments provide a picture of the gas found one billion light years away.

Similarly, when the Planck instruments observe the universe today, they see those photons that were produced 380,000 years after the Big Bang (when the emitted light was no longer absorbed by the plasma) and have travelled undisturbed for 13.8 billion years. They were emitted by the plasma contained in a very thin spherical layer that today is an enormous distance away from us, illustrated in Fig. 1. That distance is actually larger than 13.8 billion light years because, in the meantime, the universe has expanded, a phenomenon that I ignored in the preceding explanation.

The expansion of the universe has a second important effect which influences the observation of photons emitted by the cosmic gas. In fact, during their long transit these packets of electromagnetic waves are changed; the continuous expansion of space, that increased a thousandfold the dimensions of the visible universe, also stretched the peaks of the waves a thousand times further apart. In this way the photons of white light, which initially had a wavelength of little less than a micrometre, have now reached a wavelength of about a millimetre; they are *microwaves* instead of visible light. In order to 'see' them Planck's instruments are not standard cameras but sophisticated microwave detectors.

The image of Fig. 1 was obtained by observing this *microwave background radiation,* which comes from every part of the sky. The average wavelength is practically the same in every direction, and coincides with what would be measured on Earth inside a container held at 3 K, that is −270 °C.

Hence the cosmic space which separates the galaxies is traversed in every direction by a radiation that is a *relic* of a remote era; 380,000 years after the Big Bang it was at a uniform temperature of 3,000 degrees, while today it is much, much colder. The picture shows that the uniformity was however not absolute. In small regions of the sky, appearing as red spots in Fig. 1, the temperature of the background radiation exceeded the average value by a few hundred-thousandths of a degree, while in other parts – the blue spots – it was a few hundred-thousandths of a degree below the average; the intermediate temperatures are shown in the figure by different colours.

The coldest regions were also those which had the highest density; the gravitational attraction there was stronger, therefore the first stars and galaxies, which began to form several hundred million years after the Big Bang, had their origins in those darker spots. Their light required almost 14 billion years to reach the Planck space observatory.

A Time Machine

Even if our descendants were to observe the heavens with highly advanced instruments, capable of detecting electromagnetic waves of any wavelength with maximum sensitivity, they still would never succeed to 'see' what happened earlier than 380,000 years from the Big Bang.

They are prevented – and always will be prevented – by the opaqueness of the particle soup, which absorbed all photons when it was at temperatures higher than several thousand degrees. But physicists do not give up easily, and for three decades they have chosen the only possible way to overcome this difficulty: by constructing models.

These models are based on experimental studies of the reactions which took place between particles during the first 380,000 years of the life of the universe. The experiments were carried out using particle accelerators, devices in which tiny granules of matter, brought to speeds close to the velocity of light, are made to collide.

Thanks to Einstein's equivalence between mass and energy, the collisions in an accelerator enable the creation of new particles (provided that they have masses lower than the combined energy of the colliding particles) and the observation of their subsequent *decay*, which is the process that transforms them back into particles of ordinary matter.

By using a sufficiently powerful accelerator, it is possible to observe the creation and decays of particles that existed in the early universe when the temperature was higher. In other words, ever more powerful accelerators allow to reproduce in the laboratory some of the reactions among particles that happened further and further back in time. The accelerators are therefore not only subatomic microscopes and factories of unstable particles; they are also 'time machines'.

When 50 years ago, shortly after arriving at CERN, I made my first, small contribution to these studies, the energies released in collisions were of the order of one GeV, or one billion electron-volts. Today we know that such energies are characteristic of collisions which took place a *microsecond* after the Big Bang.

At the end of 2000, the group which I coordinated at the LEP (Large Electron Positron) collider accelerator at CERN in Geneva, collected data from collisions at 100 GeV. An increase of 100 times in energy allows to go further back in time by a factor $100^2 = 10,000$; thanks to LEP we could therefore understand in detail what happened just a *ten thousandth of a microsecond* after the Big Bang. (Physicists quantify such a small number by writing 10^{-10} s, which corresponds to 1 divided by 1 followed by 10 zeros.)

This information, combining what was known from experimental studies carried out at lower energies with some new and important theoretical progress, allowed the formulation of a mathematical model that explains the observed data contained in Fig. 1 that, as we have seen, goes back to 380,000 years after the Big Bang.

But we aim at learning more, at exploring times even more remote, earlier than 10^{-10} s, to search for confirmation of theoretical intuitions and to approach even closer to the instant in which everything began.

The new LHC accelerator at CERN has allowed us to make another step back in time by a factor 100, to a *millionth of a microsecond* (10^{-12} s) from the Big Bang.

From Cosmology to Medicine

More than a hundred years have passed since the discovery of X-rays, made possible by one of the most sophisticated particle accelerators available at the time; that was where modern physics began.

The following chapters retrace the principal events – which, following that discovery, have brought into being more and more powerful accelerators – and introduce the accelerator experts, who made possible the collisions of particles of larger and larger energies. We will get also to know the experimentalists, who invented, built and used the particle detectors needed to 'see' the particle collisions; and the theorists, who constructed the mathematical theories that best explain the experimental data. Accelerator experts have been instrumental in the understanding of both the matter around us and the history of the early universe. Since they are the least known of all these scientists, in this book I devote special attention to them and their inventions while I mention only cursorily the contributions of the experimentalists and I pass over most of the theorists, who usually get the largest credits.

The fascinating story of accelerators leads us up to the 2012 announcement of the discovery of the *Higgs boson* and the real frontiers of *high energy physics*, that the LHC – the extraordinary CERN apparatus – will enable to explore.

In the last two chapters we will also see that particle accelerators, invented and used for pure research in the field of fundamental physics, turn out also to be highly valuable in medicine, making possible the realisation of new techniques for diagnosis and treatment of numerous illnesses.

This is an interesting example that illustrates a very general fact; instruments developed by physicists for fundamental research have found – and, I am convinced, always will find – practical applications that go well beyond the intentions of their creators and that, perhaps in the long term, will bring benefits to everyone.

Contents

Chapter 1
The First Fifty Years

Contents

© Springer International Publishing Switzerland 2015
U. Amaldi, *Particle Accelerators: From Big Bang Physics to Hadron Therapy*,
DOI 10.1007/978-3-319-08870-9_1

1

Daniele Bergesio, TERA Foundation

It was snowing heavily when, at the beginning of 1960, I passed through the CERN entrance gate, which was then always open and practically unattended, for the first time. The road – which ran from Geneva to the village of Meyrin, where the research centre is located – was then quite narrow with little traffic. The laboratories and offices, constructed in the architectural style of the day, were neat, with aromas of wood and coffee.

Today the same road is often congested with traffic, despite being much wider and definitely improved by tramways going to the city, while on the main part of the site large and small buildings are crowded together in a disorderly way. When I arrive by car at the labs and offices used by the TERA Foundation - which was created twenty years ago to promote research into tumour therapy by means of hadron irradiation – I recognise them one by one like old friends, arranged along a network of streets whose names are distributed somewhat haphazardly.

The CERN roads are in fact named after great physicists of the past and to wander along them is to be reminded of the history of physics, jumping from one century to another, from a great discovery to an invention which changed our history.

Our story starts at the CERN main building, designed originally by the architects Peter and Rudolf Steiger, which hosts the main auditorium, post office, bank, main restaurant, newsagent, and travel agency. Leaving the restaurant behind us – from which on clear days you can glimpse views of Mont Blanc – we descend along Route Marie Curie, running along the site boundary. Route Röntgen opens on the right, close to one of the secondary entrances to CERN, where the fence ends the world of particle physics and long rows of vines extend.

A Mysterious Radiation

The physics we will talk about came into being on the evening of Friday 9 November 1895, when Mrs Röntgen found the dinner getting cold because her husband had not come up from his laboratory, located on the lower floor of the house put at his disposal by the University of Würzburg. That day Wilhelm Röntgen had actually observed a phenomenon missed by his many colleagues, who had for years used similar, but less performing, instruments; electrons, accelerated by a potential difference of some ten thousand volts, produced a new, very penetrating radiation when they struck the base of a glass tube and the positive electrode.[1]

The new invisible radiations – so mysterious that the discoverer named them 'X-rays' – were electrically neutral, hence very different from a bunch of electrons (which is negatively charged, because every electron carries an elementary negative electric charge).

The X-ray particles, as were later shown, are in fact 'packets' of electromagnetic energy similar to the *photons* which make up every beam of light. But a Röntgen photon carries a thousand times more energy than a photon of visible light and is therefore much more penetrating when it encounters matter. Mrs Röntgen could verify this when she saw the bones of her own left hand, with the ring she wore on

---

[1] The positive electrode (anode) of the tube used by Röntgen – lent by the great physicist Philipp Lenard – was special because made of platinum. This fact increased the production rate of the new radiation and contributed to the discovery.

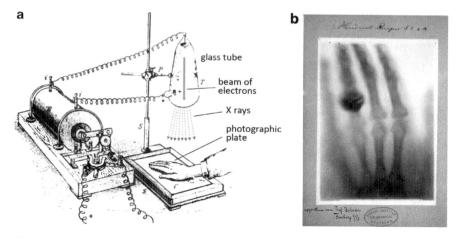

Fig. 1.1 (a) Apparatus used to make the first radiograph in history, on 22 December 1895 (Science Photo Library). (b) The flux of X rays was very small and the exposures were lengthy: Mrs Röntgen had to keep her hand immobile for 15 min (Courtesy Archive Deutsches Roentgen-Museum, Remscheid)

her finger, on a photographic plate recently developed by her husband, which is shown in Fig. 1.1.

Röntgen's Accelerator and Its Effects

Wilhelm Conrad Röntgen, born in Holland in 1845, was tall, reserved and quite shy, but with a lively character (Fig. 1.2); he and his wife were bound by a great affection, strengthened by a shared passion for sailing and the mountains. As a scientist, he preferred to work alone, constructing his instruments on his own, so that he had no real need of an assistant, and he continued in that way at the University of Würzburg, after he was appointed director of the new institute of physics at the age of 43.

On the 8 November 1895, Röntgen began a series of experiments in which a beam of electrons, accelerated by a potential difference of around 20,000 V in a glass tube evacuated of air, struck the base of the container, covered externally by black cardboard. This was therefore an *electrostatic accelerator*, in which charged particles are accelerated by a constant electric field produced between two electrodes. Under the action of this potential the electrons acquired a very high energy for that time: 20,000 eV.[2]

---

[2] 1 eV (electron-volt) is the energy which an electron acquires when it is accelerated by a voltage of 1 V.

Fig. 1.2 Wilhelm Conrad Röntgen (1845–1923) (Courtesy Archive Deutsches Roentgen-Museum, Remscheid)

In the darkness, Röntgen noticed by chance a glimmer originating from a bench a metre away, where a fluorescent sheet had been placed; on the sheet a weak luminescence appeared each time the electrons were accelerated.

Having made several trials, after a few minutes Röntgen concluded that a previously unknown radiation was being emitted from the point at which the electrons struck the glass. In the paper "On a new kind of rays", submitted on December 28 to the Proceedings of the Würzburg *Physical-Medical Society*, Röntgen described his many experiments with the following words:

> *The tube is surrounded by a fairly close-fitting shield of black paper; it is then possible to see, in a completely darkened room, that paper covered on one side with barium platino-cyanide light up with brilliant fluorescence when brought into the neighbourhood of the tube. The fluorescence is still visible at two metres distance. It is seen, therefore, that some agent is capable of penetrating a black cardboard which is quite opaque to ultra-violet light, sunlight, or arc-light. It is readily shown that all bodies possess this same transparency, but in varying degrees.*
>
> *For example the fluorescent screen will light up when placed behind a book of a thousand pages. Similarly the fluorescence shows behind two packs of cards. Thick blocks of wood are still transparent. A piece of sheet aluminium, 15 mm thick, still allowed the X-rays (as I will call the rays, for the sake of brevity) to pass, but greatly reduced the fluorescence. If the hand be held before the fluorescent screen, the shadow shows the bones darkly, with only faint outlines of the surrounding tissues. Lead 1.5 mm thick is practically opaque. The salts of the metal, either solid or in solution, behave generally as the metals themselves. The preceding experiments lead to the conclusion that the density of the bodies is the property whose variation mainly affects their permeability.* (Röntgen 1895)

The speed with which Röntgen reached these conclusions was a consequence of his skill as an experimenter and the amount of time he devoted, over a few days, to the most varied experiments; so dedicated was he that he moved his bed to the laboratory. Referring to the first time he had observed the unexpected glimmer in the darkness, a journalist enquired: "*And what did you think?*". Röntgen replied: "*I didn't think, I investigated*" (Dam 1896).

The discovery of the possibility of "photographing the invisible" had an enormous impact. Within a few weeks X-ray tubes had spread into laboratories throughout the world, as instruments for diagnosing illnesses and malformations.

On 13 January 1896 Kaiser Wilhelm II invited Röntgen to Berlin, to be informed at first hand. In the same year, more than a thousand scientific articles and 50 books on the use of the new rays appeared; already by January they were being used in Vienna to destroy a skin tumour. In popular publications the urban myth of 'X-ray glasses' was spreading, and advertisements for ladies' protective underwear appeared in an English newspaper.

In 1895 the experimental apparatus used by Wilhelm Röntgen was similar to that then available in a dozen laboratories, mostly in Europe (Fig. 1.1). However, he was the first to use such an instrument as a *fixed target accelerator,* which means to accelerate particles against a motionless matter target. His discovery marked the beginning of physics research carried out using accelerators. In 1901 Röntgen was chosen, from among twelve illustrious candidates, to receive the recently instituted Nobel Prize for physics.

Only many years later came the discovery of the value of X-rays to astronomy, when in the 1960s, with the advent of artificial satellites, it became possible to detect X-rays originating from outer space (which are normally stopped by the Earth's atmosphere); new astrophysical phenomena and heavenly bodies then became visible to us and X-ray astronomy was born.

The discovery of X-rays has thus originated two scientific and technological paths which in a hundred years, evolving in parallel and crossing innumerable times, have on the one hand changed our view of the world, in particular the cosmos; on the other, they have contributed to the survival and improvement of the quality of life of hundreds of millions of ailing persons.

I have dedicated a good part of my professional life to these two vast fields of research, which today take the name 'astroparticle physics' and 'medical radiation physics'. The first subject is described in Chaps. 4, 5, and 6 and in the epilogue, while I will discuss medical physics in Chaps. 7 and 8.

To explain both these paths I have chosen as a common thread the development of particle accelerators, which in 120 years have increased from twenty centimetres in length in Röntgen's first tube to the 27 km of the Large Hadron Collider (LHC), which accelerates protons at CERN up to an energy that is a billion times greater than those electrons of Röntgen.

This chapter recounts the first part of the history of this impressive progress.

The First Nuclear Reactions

By the beginning of the twentieth century all laboratories and very many hospitals had electron accelerators similar to Röntgen's.

The interest in using *protons* for acceleration came about in the 1920s, after the discovery in 1919 by Ernest Rutherford at the Cavendish Laboratory in Cambridge of the first nuclear reaction. Rutherford had directed a bunch of *alpha particles* (helium nuclei) - emitted by a radioactive source with an energy of 5 million electron-volts (5 MeV) – towards a target of gaseous nitrogen and observed the production of an oxygen nucleus and a proton.

The nuclear reaction which took place in the collisions is written like this, using modern notation:

$$^4_2He + {}^{13}_7N \rightarrow {}^{16}_8O + {}^1_1H$$

This formula expresses the fact that it is possible to transform a nitrogen nucleus ($^{13}_7N$) into an oxygen nucleus ($^{16}_8O$). Thus had come true, as was readily said at the time, the alchemist's dream: to transform one chemical element into another.

To understand the numbers attached to the chemical symbols of the elements in the formula, it is necessary to remember that a nucleus is made of (electrically positive) protons and (electrically neutral) neutrons; neutrons and protons are collectively called *nucleons*. They are so small that one thousand billion of them have to be put one after the other to cover 1 mm.

The hydrogen nucleus is made of a single proton; it is the simplest of all nuclei and indicated by the symbol ^1_1H. The symbol $^{13}_7N$ denotes that nitrogen is made of 7 protons and 6 neutrons, so that the total number of nucleons is 13. From the formula, one can see that the oxygen nucleus produced in the reaction is made of 16 nucleons, of which 8 are protons and 8 neutrons.

The projectile which causes the reaction is the helium nucleus, symbol ^4_2He, made of 4 nucleons, two of which are protons; it is emitted by radioactive substances, like radium and uranium, and some years earlier it had been baptised the *alpha particle* by Rutherford.

It was no accident that Rutherford made this discovery. The son of a Scottish farmer who had emigrated to New Zealand, and the fourth of 12 children, by the beginning of the twentieth century he was already noted for his studies on natural radioactivity. He had actually discovered that this was not a molecular phenomenon, but had a deeper origin from disintegration of individual atoms, and that a nucleus subject to spontaneous decay was *transmuted* into the nucleus of a different chemical element.

Rutherford was tall and robust and inspired awe, partly because of his booming voice. But despite appearances and the aura which soon surrounded him, he was a simple and generous man; among other things he encouraged the study of physics by women, something very unusual at that time.

For "his investigations into the disintegration of the elements, and the chemistry of radioactive substances" Rutherford won the 1908 Nobel Prize for chemistry. During the Nobel banquet, with his usual good humour, he claimed: "*I have dealt with many different transformations with various periods of time, but the quickest that I have met was my own transformation in one moment from a physicist to a chemist.*"

It was only 3 years later, in 1911, that Rutherford formulated the model of the atom which was destined to make him famous. He was working at Manchester University, where he was appointed in 1907 with the then highest professorial salary in the world. Under his guidance, in 1909 Hans Geiger and Ernest Marsden directed a beam of alpha particles onto a thin gold foil and observed that, while the major part of these particles crossed the target without any deviation, a few were deflected, albeit slightly, while very few others – one in ten thousand – were dispersed at angles even greater than 90°, that is backwards (Fig. 1.3).

This was an enormous surprise; as Rutherford wrote some time later:

> It was quite the most incredible event that has ever happened to me in my life. It was almost as incredible as if you fired a 15-inch shell at a piece of tissue paper and it came back and hit you. On consideration, I realized that this scattering backward must be the result of a single collision, and when I made calculations I saw that it was impossible to get anything of that order of magnitude unless you took a system in which the greater part of the mass of the atom was concentrated in a minute nucleus. It was then that I had the idea of an atom with a minute massive centre, carrying a charge. (Andrade 1964)

Thus was born the Rutherford atomic model, constructed like a miniature planetary system, whose Sun is the positively charged nucleus, constituted, as was discovered in subsequent years, of protons and neutrons, and whose planets are the negative electrons. All the mass is in the nucleus, because although the number of electrons is equal to the number of protons, the mass of an electron is 2,000 times smaller than the mass of a proton (and of a neutron, too).

In Rutherford's experiment almost all the alpha particles passed through empty space and the sparse electronic clouds between the nuclei without changing direction, but a few collided with the dense nuclear charge, being deflected and, very rarely, turned back.

The experiment of Rutherford is the prototype of many *fixed target* experiments still carried out today with particle accelerators. In this type of experiment, a particle beam is directed against a matter target, traversing which the fast particles strike nuclei of the atoms. Observing the *final state* particles leaving the target, it is possible to study the phenomena which occur during the brief moments of the collision.

The Cockcroft-Walton Accelerator

In the 1919 experiments of Rutherford, the alpha particles, used as projectiles to transmute nitrogen into oxygen, were emitted by a radioactive source. After collimating them, or 'squeezing' to form a narrow beam by means of an electric field, they were directed against a target (Fig. 1.4).

In 1927, in his annual address given as President of the Royal Society, Rutherford described the objective of making available for experimentation a copious *artificial* source of atoms and electrons, with energies far greater than that of naturally emitted alpha particles, to probe and disintegrate nuclei more effectively.

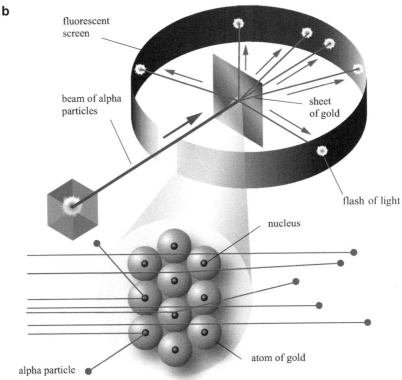

Fig. 1.3 A few of the alpha particles which collide with the atoms of a gold foil are deflected through large angles. Rutherford, in the photo with Hans Geiger, deduced that at the centre of the atom there is a minute, positively charged nucleus in which almost all the mass is concentrated ((**a**) © Bettmann, CORBIS)

Fig. 1.4 In the nuclear transmutation observed in 1919 by Rutherford (at the centre of the photograph) the 4 nucleons of the helium nucleus (alpha particle) and the 13 nucleons of the nitrogen nucleus are rearranged to form nuclei of hydrogen (1 nucleon) and oxygen (16 nucleons). In 1932 his students John Cockcroft (on the right in the photo) and Ernest Walton observed the first nuclear transformation produced by artificially accelerated protons, instead of particles emitted naturally by a radioactive substance ((**a**) Pictorial Press Ltd/Alamy)

There is still much work to be done before we can hope to produce streams of atoms and electrons of a much higher individual energy than the alpha-particle spontaneously liberated from radioactive bodies. The alpha-particle from radium C^3 is initially expelled with an energy of about 8 million electron volts. So far the alpha-particle has the greatest individual energy of any particle known to science, and for this reason it has been invaluable in exploring the inner structure of the atom and giving us important data on

---

[3] At Rutherford's time the decay products of radium were called A, B, C... Later it became clear that they were different chemical elements; radium C is bismuth ^{214}Bi.

the magnitude of the deflecting field in the neighbourhood of atomic nuclei and of the dimensions of the nuclei. In the case of some of the lighter atoms, the alpha-particle has sufficient energy to penetrate deeply into the nucleus and to cause its disintegration.

It would be of great interest if it were possible in laboratory experiments to have a supply of electrons and atoms of matter in general, of which the individual energy of motion is greater even than that of the alpha-particle. This would open up an extraordinary new field of investigation that could not fail to give us information of great value, not only in the constitution and stability of atomic nuclei but also in many other directions, (Rutherford 1928)

Not surprisingly then, at the end of 1928 John Cockcroft, Rutherford's student, wrote to the University of Cambridge with a request for £1,000, to construct a new type of electrostatic accelerator in the Cavendish Laboratory. The request was accepted and for 3 years Cockcroft worked with Ernest Walton, another student of Rutherford, on the development of different types of electrostatic accelerators. Finally in 1932, only four years later, they constructed the first 'Cockcroft-Walton' accelerator and were able to observe the first nuclear transformations produced by a beam of artificially accelerated protons.

To contrast the difference with modern times, in which research in physics is the result of huge international collaborations lasting for many years, it is enough to cite the example of the CERN Large Hadron Collider. The first outline of the LHC concept was actually presented, accompanied by a thick volume of design studies, by the Italian engineer Giorgio Brianti and his team at a conference held in Lausanne in 1984. The building of the LHC was approved in 1994 with an investment of around five billion euros and the first collisions occurred in 2009, 25 years after the original proposal!

Cockcroft's 1928 proposal foresaw the 'artificial' acceleration of hydrogen nuclei ^1_1H (i.e. protons) against a target, to produce a nuclear reaction similar to those of Rutherford. The problem was to establish how much energy would be required by a hydrogen nucleus to penetrate a target nucleus and produce a reaction. As the target nucleus is positively charged, it therefore repels the approaching proton.

Cockcroft proposed using as a target a light nucleus – lithium, made of three protons and four neutrons – to reduce the electrical repulsion of the hydrogen nucleus. However, according to classical physics the minimum energy necessary is still extremely high even in this case; to produce the reaction the incident proton should have an energy equivalent to at least several million electron-volts, a value practically unattainable in Cockcroft's time. The new quantum physics, worked out the previous year (1927), came to his aid.

From the perspective of the new theory, a proton can penetrate a nuclear target even when its energy is much below the required level, as a consequence of energetic fluctuations made possible – as we will see shortly – by the 'uncertainty principle', provided the collision takes place in a sufficiently short time. It is as if, when bouncing a ping-pong ball against a concrete wall, there were to exist a finite chance that the ball would pass through the wall, provided the traversal time is extremely short (this phenomenon is called 'tunnelling').

Rutherford suggested these theoretical arguments and Cockcroft estimated that a potential difference of 200,000 V would be sufficient, and therefore an energy of 200,000 eV (because a proton has the same charge as an electron, but positive), for a

Fig. 1.5 Ernest Walton shown sitting in the small cabin where the observations were made (Cavendish Laboratory – Cambridge University)

proton to have a non-zero probability of penetrating a lithium nucleus. The reasoning was correct and in 1932 the Cockroft-Walton accelerator became the first particle accelerator producing a nuclear reaction. For their experimental achievements, the two researchers received the 1951 Nobel Prize for physics (Fig. 1.5).

It is interesting to note that in this experiment the energy was actually far smaller than the one with alpha particles. However the discovery was made possible because a (singly charged) proton is repelled by the target nucleus four times less than a (doubly charged) alpha particle and the uncertainty principle gave a helping hand.

Mass and Energy in the Subatomic World

In the 1920s the small world of university physics had developed a common body of knowledge but, in research, was divided into two distinct communities dedicated to the understanding of very different physical systems: on one hand the nucleus, on the other the atom.

The shared foundations comprised the theory of relativity and the recognition that energy is carried by light, and other electromagnetic waves, in the form of indivisible packets of energy which – only since 1926 – had been given the name *photons*.

The principal consequence of the theory of special relativity, formulated by Einstein in 1905, is the *equivalence of mass and energy*, which is expressed by the equation $E = mc^2$, famous enough to be the only formula allowed to appear in daily newspapers. It expresses the fact that a motionless body of mass m has an energy, which can be obtained by multiplying its mass by the square of the speed of light c, which is equal to 300,000 km/s and is the *maximum speed* at which any signal can travel. The product mc^2 is the 'rest energy' of the body of mass m, that means the energy of the body when it is stationary. When the body – for example a particle – is in motion, its *total* energy is larger than its rest energy. By subtracting from the total energy of a moving particle its rest energy, one obtains the energy due to its motion, which is referred to by the name 'kinetic energy'.

In the collisions produced by particle accelerators Einstein's formula is essential because when mass disappears the corresponding amount of energy has to appear, and vice versa.

The value of c^2 is so large that to produce 3,000 kilowatt hours (the electrical energy which an average family uses in a year) it would be sufficient to annihilate one hundredth of a milligram of matter.

Given the equivalence $E = mc^2$, physicists find it convenient to measure both energies and masses with the same unit: the GeV, or 'gigaelectron-volt', which is a billion electron-volts (or in MeV – equal to one thousandth of a GeV). To measure the mass of a particle in GeV essentially means to call 'mass' the product of m and the square of the speed of light, i.e. its rest energy.

1 GeV is roughly equal to the energy required to raise 2 mg of matter a hundredth of a millimetre from the earth's surface. It is a very small quantity in our daily experience but very large when associated with a single atomic particle; it is enough to realise that there are about 10^{21} atoms in these 2 mg of matter [4]!

For our purposes it is preferable to keep in mind the fact that 1 GeV is the mass or, equivalently, the rest energy of one proton and is, therefore, a measurement unit adapted to the description of phenomena of the subatomic world (in reality the mass of the proton is slightly less, 0.938 GeV, but in what follows we will neglect this small difference).

Instead an electron has a mass of 0.0005 GeV (0.5 MeV), 2,000 times smaller than that of the proton. This means that, because energy can be transformed into mass, when an electron is created in a collision between subatomic particles, it is necessary to spend at least 0.0005 GeV of energy, i.e. 0.5 MeV.

---

[4] 10^{21} is a very large number equal to a 1 followed by 21 zeros. To express very small numbers, instead, negative powers are used: 10^{-13} equals 1 divided by 1 followed by 13 zeros.

In reality, as we will see in the next section, the energy which disappears is double that, equal to 0.001 GeV, because an electron must always be created with its 'antiparticle', an antielectron, which has the same mass but positive, instead of negative, charge (law of conservation of electric charge). Even more energy has to disappear if the newly created electron and the antielectron move, thus possessing a certain amount of kinetic energy.

Also a proton has to be created always together with its antiparticle, an antiproton, which has identical mass but negative electric charge. In such a reaction an energy equal to at least the double the proton mass has to disappear, as we shall better see in the following.

In the 1920s physicists used the theory of relativity and the mass-energy equivalence but they did not know of the existence of antielectrons and antiprotons. They knew very well, however, that the mass (or rest energy) of a photon – or 'quantum' of the electromagnetic field – is *zero*. The reason is simple: no massive body can move at the speed of light, because this speed is reachable only by providing an infinite amount of energy. Since a photon, a packet of luminous energy, travels at the speed of light, its mass can only be zero. It exists as it transports energy from the place where it is produced to the place where it is absorbed, but the transported energy is pure kinetic energy without rest energy. In other words, a photon is never at rest.

Heisenberg's Uncertainty Principle

Building on the foundations of the shared knowledge described in the previous section, towards the end of the 1920s nuclear physics was mainly experimental; the unquestioned leader of experimental nuclear physics was Ernest Rutherford. On other fronts theoretical physicists tried to understand the ambiguous behaviour of electrons and photons, which sometimes acted like particles and at other times like waves. There were many protagonists in this theoretical research, but everyone watched what was said and done by the Danish physicist Niels Bohr, professor in Copenhagen after 4 years spent working with Rutherford in Manchester. Bohr inspired and led European theoretical physicists, while he published few scientific papers of his own because he was so self-critical (Fig. 1.6).

The influence of these two giants of physics is recalled in the name 'Bohr-Rutherford model' which still today designates the structure of an atom made of a tiny central nucleus containing positive protons (discovered by Rutherford) around which circulate the very light negative electrons following only a few permitted orbits (discovered by Bohr).

In 1927 the young German physicist Werner Heisenberg, just 26 years old, made a critical analysis of the concepts of position, time and energy applied to subatomic particles (Fig. 1.6). This led him to the *uncertainty principle*, which takes account of the dual corpuscle and wavelike nature of all particles, without using the complicated quantum physics equations that Bohr, Heisenberg himself and many of their colleagues were then finalising.

When a collision creates new particles, the sum of their total energies is always exactly equal to the energy that disappears in the collision. This means that the principle of conservation of energy, according to which energy is neither created nor destroyed, also holds in the atomic and subatomic world. But Heisenberg argued that in the subatomic world there are exceptions that change everything.

The most striking form of this principle maintains that, in any physical system, conservation of energy can be violated *provided such a violation happens for a sufficiently short time*. This possibility is a consequence of the 'non-deterministic' nature of atomic and subatomic phenomena; for the very short time interval in question, energy is in reality undetermined and therefore energy conservation is not really violated, only evaded.

Quantitatively the uncertainty principle maintains that the larger the energy E that appears, violating the conservation law, the shorter the time t for which the violation can last, eluding even the most careful observation. Heisenberg's principle can thus be represented by a formula with inverse proportionality:

$$t = h/E$$

where h is a quantity now known as 'Planck's constant'.[5] If this constant were to be zero, the avoidance of the law of energy conservation would be impossible, because the duration for which energy conservation could be evaded would also always be zero. Numerically however, h is different from zero, while having an extremely small value, equal to about 10^{-34} J · s (joules multiplied by seconds); in terms of energies and times of the macroscopic world, h is therefore represented by the number 1 divided by a 1 followed by 34 zeroes.

Since the constant h is miniscule, the uncertainty principle does not have visible effects in the macroscopic world, but begins to make itself felt when descending to the level of molecules and atoms, where the energies and times in play are so small.

As I already explained, in the world of particles the natural unit of energy measurement is 1 GeV. What is the best unit to measure subatomic *time*? Certainly not the second, which is the very human time interval that separates roughly two beats of our heart.

Since the diameter of a proton is 10^{-15} m (a length that physicists call a 'fermi') it is natural to choose as a unit the time that light, moving at 3×10^8 m/s, requires to travel exactly 1 fm. Here I will call this tiny interval a 'heisenberg'– equal to 3×10^{-24} s – which is, in a certain sense, the heartbeat of a proton.

Using this unusual unit and measuring the energy E in GeV, the uncertainty principle takes the form:

$$t = 0.2/E$$

---

[5] Physicists call 'reduced Planck constant' the quantity h appearing in this equation.

The principle now states that the energy of a physical system can, at a certain instant, augment its energy by $E = 1$ GeV, but only for a time shorter than $t = 0.2/1 = 0.2$ heisenberg, or $0.2 \times 3 \times 10^{-34}$ s $= 6 \times 10^{-35}$ s. Metaphorically it is as if the system were to take a loan of the energy E from the 'Heisenberg bank', but the greater the amount of energy borrowed the earlier the loan matures.

In 1928 the Russian-Ukrainian theoretical physicist George Gamow used the uncertainty principle to explain the emission of alpha particles by heavy nuclei, such as uranium and radium. Gamow, born in Odessa, studied in Leningrad and then worked for some time with both Bohr and Rutherford before going back to the Soviet Union. He was an extremely bright and extrovert man and made important contributions to subatomic physics, to the understanding of the nuclear reactions taking place in stars and to the Big Bang theory by predicting, together with Ralph Alpher, the existence of the relic radiation, so well photographed by the Planck space observatory (Fig. 1, Prologue). Twice with his wife he tried to escape by kayak from the Soviet Union – through the Black Sea and Norway – until, on the occasion of a conference held in Brussels in 1933, he finally managed to leave the country and emigrate to the United States.

In the 1940s, to explain modern physics – that is relativity and quantum physics – to the general public George Gamow invented an amusing character, Mr Tompkins, who partly from enthusiasm and partly because he's in love with the daughter of a physics professor, becomes aware of the most surprising aspects of the latest scientific research. The titles of his books – still read today and more famous than his fundamental scientific papers – are "Mr. Tompkins in Wonderland" and "Mr. Tompkins Explores the Atom".

Let us now go back to 1928. Gamow suggested the argument to Rutherford, and Heisenberg's principle was then used by Cockcroft in the planning of his

Fig. 1.7 Gamow writes
that – due to the tunnelling
effect – it can happen that
*"a car locked safely into a
garage, leaks out just as a
good old ghost of the middle
ages"* (Gamow 1993)

accelerator since they were convinced that, from time to time, a proton, accelerated
to only 0.0002 GeV (0.2 MeV), could penetrate a lithium nucleus and split it in two,
producing two helium nuclei.

The repulsive force between the positive charge of the proton (^1_1H) and that of
the lithium nucleus (^7_3Li) constitutes a barrier which prevents the proton from
entering into the nucleus, at least until its energy exceeds several MeV. If energy
were rigorously conserved, the reaction could never occur at 0.2 MeV. Fortunately
the time needed to penetrate the barrier, which can be calculated knowing the speed
of the proton and the size of the nucleus, is so short that, thanks to the uncertainty
principle, this limit can be avoided; so occasionally a proton does penetrate the
nuclear target (as if there was an hidden 'tunnel') and causes the transmutation
observed by Cockroft and Walton in 1932.

This remarkable quantum phenomenon is jokingly represented in the drawing of
Fig. 1.7, which appears in one of Gamow's popular science books. Mr Tompkins
imagines a car suddenly appearing in a living room; thanks to the uncertainty
principle, the car penetrates the barrier dividing the garage from the house. Fortunately, the probability of such an event is completely negligible, because Planck's
constant is an extremely tiny number at the ordinary, macroscopic world scale.

The Positron and the Antiparticles

For physics, 1932 was a very eventful year. As well as the first reactions produced by artificially accelerated protons, it saw the discovery of the neutron by James Chadwick (another pupil of Rutherford), the acceleration of protons up to a million electron-volts with Ernest Lawrence's cyclotron (which we will see later in this chapter) and the discovery of the *positron* (the anti-electron or positive electron) by the American physicist Carl Anderson. One could say that particle physics was really born in 1932 with Anderson's observation of the tracks left by a positron in a cloud chamber.[6]

The discovery was possible because the universe has always contained 'natural' accelerators which are much more powerful than those we can construct and, even better, cost nothing. The Earth is bombarded constantly by a flux of *cosmic rays*, which we now know to be produced by super-massive black holes and by residues of exploding stars that we call supernovae. 90 % of these cosmic ray particles are nuclei of hydrogen, i.e. protons, and most of the remaining 10 % are helium nuclei (alpha particles).

Cosmic rays were discovered in 1911 by the Austrian physicist Victor Hess, who made a series of balloon ascents to take measurements of radiations in the atmosphere. For the first time in history this phenomenon, which has been taking place continuously for billions of years, was brought to the attention of the general public. In 1936 Hess shared the Physics Nobel Prize with Carl Anderson (Fig. 1.8).

Since the energies are extremely high, in the collisions with nuclei in the atmosphere, new particles – which mostly decay in flight – are created, as shown schematically in Fig. 1.9a. The symbols e^+ and e^- refer to positrons and electrons. The 'gamma' (γ) particles are high energy photons, similar to packets of luminous energy but which carry quantities of electromagnetic energy millions or billions of times larger. The particles labelled π, μ and ν will be discussed later.

If one wants to observe the atmospheric interactions of protons originating from intergalactic space it is necessary to ascend to high altitudes. At low elevations secondary particles are observed, especially photons.

In 1932 Anderson observed that photons produced by cosmic rays sometimes, when colliding with matter, create pairs of particles which have exactly the same mass but opposite electric charges: an electron, indistinguishable from an atomic electron, and the already mentioned antielectron, or 'positron'.

In Fig. 1.9b the cloud chamber is immersed in a magnetic field orthogonal to the plane of the paper. Two particles in the lower part of the picture can be recognised to have opposite charges because they are bent in opposite directions by the magnetic field. In this phenomenon the energy of the incident photon disappears, and the mass of the electron and positron – together with their kinetic energy – is produced in agreement with the Einstein relation $m = E/c^2$.

---

[6] A cloud chamber contains water vapour in a special unstable state called 'supersaturated'. If a charged particle crosses it, the ionised atoms become condensation centres for miniscule drops of water, causing 'vapour trails' which make the particle trajectories visible.

Fig. 1.8 Victor Hess discovered that at an altitude of 5,000 m the intensity of the local radiation is twice that at sea level and concluded that Earth is continuously bombarded by what were immediately called 'Hess rays' and later became known as 'cosmic rays' (Courtesy VF Hess Society, Schloss Pöllau/Austria)

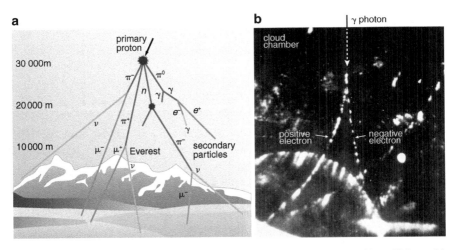

Fig. 1.9 (**a**) High energy photons (symbol γ) are among the particles produced by collisions with cosmic rays. (**b**) The creation of a negative electron and a positive electron, observed at the beginning of the 1930s in a cloud chamber in a magnetic field (I. Curie e F. Joliot, Science Photo Library)

The excitement caused by Anderson's discovery was enormous, although the idea that in certain processes new particles could be generated was not new to physicists. Photon emission by neon atoms contained in a luminescent tube was actually well known. When an electric discharge passes through a rarefied gas, many gas atoms acquire energy which, almost immediately, is reemitted in the form of photons with a characteristically blue colour. Each atom found in a highly energetic state emits a photon. This is created – at the expense of energy from the atom – in the same instant that the atom passes from a state of higher energy to a lower energy one.

The creation of a *single* neutral particle was therefore already known, but not a *pair* of particles of opposite charge and equal mass. The phenomenon is described by a new type of transmutation reaction, in which the energy of a γ particle (which is neutral, with no mass) is transformed into two new particles, an electron and a positron (which have equal masses and are charged):

$$\gamma + \text{nucleus} \rightarrow e^- + e^+ + \text{nucleus}$$

Since the photon is electrically neutral, by the law of conservation of electric charge an electron and a positron must always be created in pairs.

In this process an energetic γ photon interacts with the charge of an atomic nucleus and disappears, using part of its energy to create the mass of the electron e^- and positron e^+. Since the mass of an electron corresponds to an energy of 0.5 MeV, the photon must have an energy greater than 1 MeV to create the e^+-e^- pair, as I said when I discussed the relationship between mass and energy in the subatomic world. The remainder of the photon energy exceeding 1 MeV becomes kinetic energy of the newly created particles.

The Dirac Equation and Antimatter

Two years before Anderson's discovery, the existence of the positron had been predicted by the British physicist Paul Dirac, who in 1928 had formulated a quantum theory capable of describing the behaviour of an electron moving at close to the speed of light. The fundamental equation in Dirac's theory combined special relativity with quantum physics and predicted the existence of a particle with the same mass as an electron, but opposite charge.

The concept of 'antielectron' was therefore already known to European physicists. However Anderson, who worked in California, was not aware of Dirac's equation when he discovered the positron and announced it in fall 1932. In February 1933 Patrick Blackett and Giuseppe Occhialini – who were working together in Cambridge since 1932 – announced the results of their study of 700 photos taken with the Cavendish Cloud chamber: 14 tracks showed a positron track. Unlike Anderson, Blackett and Occhialini explicitly associated the observation of the new particle with the antielectron predicted by Dirac's quantum theory. Their observations not only confirmed Anderson's discovery, but also pointed out the existence of 'showers' of electrons and positrons arriving on Earth in equal numbers.

Subsequently the opposite phenomenon to electron-positron *pair production* was observed: a positron that, crossing a slab of matter, is slowed by collisions with the atomic electrons, sometimes encounters one of them so closely that it produces an *annihilation*. The positron and electron disappear with their charges and their energy and produce two (and sometimes three) photons. Therefore the electromagnetic energy of a photon can be transformed into a particle of matter and a particle of 'antimatter' and, vice versa, matter and antimatter can dissolve into electromagnetic energy: a pure manifestation of Einstein's formula in action.

The revolutionary aspect of Dirac's theory and its experimental confirmation touch on the very meaning of the adjective 'elementary'. Heisenberg expressed it in a very clear way:

> *I believe that the discovery of particles and antiparticles by Dirac has changed our whole outlook on atomic physics. As soon as one knows that one can create pairs, then one has to consider an elementary particle as a compound system; because virtually it could be this particle plus a pair of particles plus two pairs and so on, and so all of a sudden the whole idea of elementary particles has changed. Up to then I think every physicist had thought of the elementary particles along the lines of the philosophy of Democritus, namely by considering these elementary particles as unchangeable units which are just given in nature and are just always the same thing, they never can be transmuted into anything else. After Dirac's discovery everything looked different, because now one could ask, why should a proton be only a proton, why should a proton not sometimes be a proton plus a pair of electron and positron and so on?* (Heisenberg 1969)

For this reason it is preferred to talk of 'fundamental', rather than elementary, subatomic particles.

In summary, an antielectron has mass and properties identical to those of an electron, but its electric charge is positive instead of negative. Symmetrically, an antiproton, the antiparticle of the proton, is electrically negative while the proton is positive; but both have a mass about 2,000 times larger than an electron or a positron.

Today we know that all the electrically charged particles have their own antiparticles. For the neutral particles things are more complicated: some have an antiparticle while others, like photons, are at the same time both particle and antiparticle.

When a particle and its antiparticle encounter one another, they 'annihilate' releasing energy. Where does this energy come from? Evidently from the mass and the kinetic energy of the particle and antiparticle which disappear.

Conversely, with a sufficient quantity of energy available, in the right conditions particle-antiparticle pairs can be created. As already explained, to create the mass of a proton 1 GeV of energy is enough, but to produce it in practice at least 2 GeV are required, because it is necessary to create simultaneously the mass of a proton and that of an antiproton. If in the process more than 2 GeV are used, the rest is converted into kinetic energy of the newly created proton and antiproton.

In the CERN LEP accelerator, in use until 2000, electrons and positrons of about 100 GeV kinetic energy annihilated each other, thus releasing a total energy of

200 GeV. This energy in its turn caused the creation of particle-antiparticle pairs that, in a cascade, produced even more, permitting the discovery of new particles.

Today's LHC accelerator, in which two beams of protons collide, beats this record by over ten times, releasing an energy equal to thousands of GeV, i.e. several TeV (*tera*electron-volts, where tera $= 10^{12}$).

The Invention of the Cyclotron

Hardly had Cockcroft and Walton produced the first nuclear reactions with artificially accelerated particles in 1932, than many other research groups reproduced the same result. At Berkeley, in the United States, Ernest Orlando Lawrence had already 3 years earlier begun to construct his series of cyclotrons, a new type of accelerator, which used a magnetic field to bend the trajectories of the accelerated particles. He wrote:

> We were busy with further improvements of the apparatus (the cyclotron) to produce larger currents at higher voltages when we received word of the discovery by Cockcroft and Walton. We were overjoyed with the news, for it constituted definite assurance that the acceleration of charged particles to high speeds was a worthwhile endeavour. (Lawrence 1951)

Three years earlier, on 1 April 1929, Lawrence – just appointed associated professor in Berkeley – was scanning an article in German by the Norwegian physicist Rolf Wideröe in the university library. He did not know the language but studied the illustrations (in the upper part of Fig. 1.10) with great attention. That very afternoon Lawrence had the happy idea of *adding a magnetic field* to the linear accelerator proposed by Wideröe.

In a magnetic field the protons do not follow a straight line, as in the electrostatic accelerators, but travel along a circular trajectory; furthermore the radius of the trajectory of the protons grows with the increase in their energy.

The idea was to make groups of protons circulate many times inside two hollow electrodes, one as in shape C and the other like D, placed inside the poles of a magnet (Fig. 1.11). The protons are accelerated each half turn by an alternating voltage applied to the electrodes. Combining the accelerating action of the electric field with the bending effect of the magnetic field, the particle bunches are accelerated along a spiral trajectory, which allows to reach higher energies with a smaller apparatus and lower voltages at the electrodes with respect to electrostatic accelerators.

The principle of the cyclotron is illustrated in the lower diagram of Fig. 1.11. The protons are 'injected' with very low energy at the centre of the two electrodes. The source of protons is hydrogen gas, in which an electric discharge detaches the electrons from the nuclei; the protons are these 'bare nuclei'.

Near the centre of the magnet, all the protons, which pass at the right moment into the space separating electrode D from C, acquire an energy E when D is

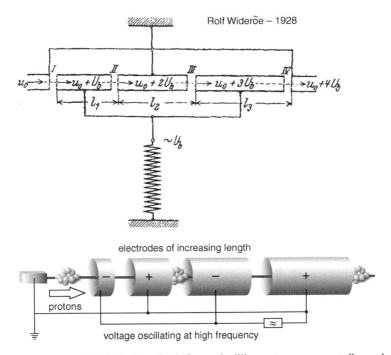

Fig. 1.10 The scheme of Wideröe: 'bunches' of several million protons are repeatedly accelerated by the potential difference applied between 'drift tubes'. The length of the tubes increases progressively along the way, to compensate for the increase in the velocity of the protons and the corresponding reduction in transit time ((**a**) Courtesy Lawrence Berkeley National Laboratory)

positive and C negative. In the magnetic field their trajectory is highly curved, because these protons have little energy.

When the bunch of protons, which has a length of a few centimetres, arrives again at the space which separates C from D after having travelled half a turn, the alternating voltage applied by the radiofrequency source has changed direction: now C is positive and D negative. All the protons in the bunch therefore receive a second amount of energy E. Furnished with kinetic energy $2E$, they travel along a trajectory with a slightly larger radius. After another half turn, they again arrive at the space separating D from C, where they receive another boost of energy E, reaching, all together, a kinetic energy of $3E$.

After hundreds of turns the trajectory of the bunch of protons is perturbed by a deflector, as shown schematically in the lower part of Fig. 1.11, and the protons continue in a straight line out of the magnetic field of the cyclotron, thus forming what is called a 'beam' of particles, which is generally used to bombard a fixed target. So that the protons are not lost as a result of collisions with air atoms, the source, the two electrodes and the extracted beam are contained in a vacuum chamber, from which air has been removed.

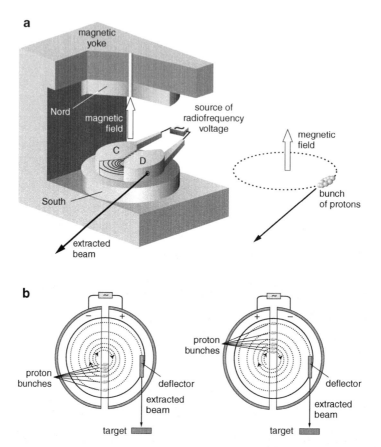

Fig. 1.11 (**a**) The circular iron poles (connected by a 'yoke', also of iron, cut away in the diagram above) produce a vertical magnetic field that bends the trajectories of the protons. (**b**) With each transit between electrodes C and D the protons acquire extra energy and therefore circulate in the magnetic field along a trajectory of increasing radius of curvature that takes the form of a spiral

Lawrence's Many Cyclotrons

That famous 1 April, in the Berkeley library, Lawrence sketched out some calculations to confirm the practicality of his intuition. He immediately found out that, as the energy grew, the increase in the trajectory radius should exactly compensate for the increase in speed: the proton bunches would always travel through each half turn in exactly the *same* time interval and remain in synchronism with the accelerating electric field.

The simple equations made him realise that, as if by magic, there should be perfect synchronisation between the regular alternation of the voltage between the electrodes and the successive transits by the bunch of particles.

Lawrence was a young 27-year-old experimental scientist, full of energy and enthusiasm. As soon as he left the library, he could not stop himself revealing his

Fig. 1.12 Ernest Lawrence – *right* in figure (**b**) and second from the left in (**c**) – constructed a series of cyclotrons, each larger and more powerful ((**a**) Lawrence Berkeley Lab/Science Photo Library; (**b** and **c**) Courtesy Lawrence Berkeley National Laboratory)

idea to a professor he ran into on the campus, who asked him: "*But what will you do with it?*". The immediate answer was "*I will bombard and split atoms!*" (Childs 1968). Lawrence was still more excited the following afternoon when, meeting another more senior professor, he cried "*I will be famous!*" (Childs 1968). He was not wrong: in 1939, among many other honours, he was awarded the Nobel Prize for physics.

As the drawings in the lower part of Fig. 1.11 show, in the cyclotron at any instant there are bunches of protons of all energies, which rotate in synchronisation along circular paths of different radii.

According to the equations written down by Lawrence that day, these bunches arrive together at the space between the two electrodes, exactly in time to receive another accelerating kick. The radiofrequency source must produce an alternating voltage with the correct frequency to ensure this synchronisation.

The first cyclotron (Fig. 1.12a) was as big as a hand, while a Cockcroft-Walton accelerator of the same energy would have been more than a metre long!

The magnetic poles had a diameter of 4 in., about 10 cm, and the alternating potential was only 1,800 V; in about 20 turns the protons reached 80,000 eV.

The second cyclotron, built together with doctoral student Stan Livingston, used a magnetic yoke of 80 t, constructed for other reasons by the Federal Telegraph Company; the cyclotron was initially 9 in. in diameter, but was soon increased to an apparatus of 11 in. (28 cm).

Protons were accelerated up to 1.2 MeV and in 1932 a scientific paper written by Lawrence and Livingston appeared. The title was "The Production of High Speed Light Ions Without the Use of High Voltages", fully justified because, by making the proton bunches circulate 150 times, they exceeded a MeV in kinetic energy by applying an alternating potential of only 4,000 V.

To continue further along the same path Lawrence obtained financial support from public foundations and private sources; as we will see in the final chapter, the new accelerators produced radioisotopes useful for medicine – what is now called 'nuclear medicine' – and were also soon employed for cancer therapy.

Lawrence was more interested in the technological challenges of ever larger diameter cyclotrons, and the medical use of his invention, than in the results of nuclear physics experiments, as acknowledged by Stan Livingston who, years later, wrote:

> The accelerator scientists saw the need for high-energy physics, but did not have any great vision of a new field. They had their own motivations external to the general development of nuclear physics and Lawrence was not particularly impressed with the growth of nuclear physics, but he enjoyed the technical success. (Wiener and Hart 1972)

He anyway gave generous access to his own machines and help to copy them wherever needed. He obtained a patent on his invention, but he never wanted to make money from it.

Effects of Relativity

In 1940, at the Radiation Laboratory in Berkeley, Lawrence began the construction of an enormous magnet of 184 in. (around 4.7 m) and 4,500 t for his latest accelerating machine, which would reach 100 MeV, later increased to 200 MeV.

Launching the project, the American physicist was not too concerned by the fact that at kinetic energies larger than about 10–20 MeV the protons circulating in a cyclotron start to behave differently; according to Einstein's relativity they require significantly more time to travel the half-circumference, so losing synchronisation with the alternation of the voltage on the electrodes. As we will see in the next paragraph, beyond these energies Lawrence's machine – as originally planned – could never have worked, because of the presence of these 'relativistic effects'.

To explain the situation I refer to Fig. 1.13 that shows the 'relativistic factor' by which the particle mass is multiplied to obtain its *total* energy. This factor – expressed by the formula appearing in the figure – depends on the ratio β of the

Fig. 1.13 The *relativistic factor* by which the mass is multiplied to obtain the *total* energy when a body moves at a fraction $\beta = v/c$ of the speed of light $c = 300,000$ km/s. When β approaches 100 %, the relativistic factor increases indefinitely

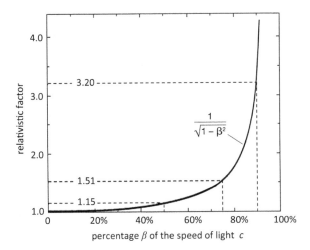

Table 1.1 Energy and kinetic energy of a proton as computed from its speed. The proton is assumed to have a mass of 1 GeV

Speed in km/s	Speed β as percentage of the speed of light (%)	Relativistic factor	Total energy of the proton in GeV	(Kinetic) energy of the proton in GeV
60,000	20	1.02	1.020	0.020
125,000	42	1.10	1.100	0.100
150,000	50	1.15	1.150	0.150
166,000	55	1.20	1.200	0.200
225,000	75	1.51	1.510	0.510
270,000	90	3.20	3.200	2.200
297,000	99	7.09	7.090	6.090
299,250	99.7	14.2	14.200	13.200
299,970	99.99	70.7	70.700	69.700

particle speed to the speed of light and allows the calculation of the total energy of the particle for any speed and, vice versa, of the speed once the mass and the total energy are known.(Note that, in ordinary life, speeds are always much lower than 0.001 % of the speed of light and the relativistic factor is practically equal to 1)

Table 1.1 lists some typical values of the relativistic factor, calculated for a proton which moves at the speed shown in the first column. In the last column the kinetic energy E of the proton is calculated from the difference between its total energy E_{tot} and its mass M (measured in GeV, i.e. its rest energy). In common parlance the specification 'kinetic' is often dropped and the term 'energy' is used instead of 'kinetic energy'; for instance, as we shall see in Chap. 8, the protons used in the treatment of deep seated tumours have energies of 0.2 GeV (200 MeV) and move – as indicated in the table – at 55 % of the speed of light.

The figure and the table show that, if a particle moves at 50 % of the speed of light, its kinetic energy is 0.15 GeV, 15 % larger than its mass. The energy is 51 % larger than the mass if the speed reaches 75 % of the speed of light.

At larger speeds the kinetic energy is greater than the mass; so, for example, a proton which travels at 270,000 km/s has a (total) energy of 3.2 GeV and a kinetic energy equal to 2.2 GeV.

The fact that the energy of an accelerated particle becomes continuously larger than its mass has a great impact on the functioning of particle accelerators, because in a magnetic field the trajectory is determined by its inertia (i.e. by its reluctance to be deflected from a straight path) which is given *not* by the mass *but* by the total energy of the particle; it is the total energy that fixes the radius of the particle's circular path in a magnetic field.

In summary, for slow particles the kinetic energy is much smaller than the mass, the relativistic factor is 1 and the mass, which is equal to the total energy, measures the inertia. For fast moving particles the total energy plays the role that mass has for slow particles and many physicists like to call it 'mass in motion'.[7]

The Phase Stability Principle

As already mentioned, the construction of the giant magnet for Lawrence's new cyclotron began before he found a means of overcoming the obstacle of the relativistic increase of inertia. However, after a year the project was interrupted by the war.

On 7 December 1941 the American Pacific fleet was almost completely destroyed by a surprise Japanese attack at Pearl Harbour, in Hawaii. Ten days later the OSRD (the US government Office of Scientific Research and Development) assigned to Lawrence's Radiation Laboratory, as part of the Manhattan Project, a sizeable amount of money to develop a method, reproducible on a large scale, for separating the fissile isotope uranium-235 from uranium-238.

Lawrence then hastened the construction of the 184 in. magnet and, within only six months, began to realise the first prototypes of 'magnetic separators', called *calutrons* (after 'California University'), which then were inserted between the two poles of the magnet originally intended for the new cyclotron. These machines were then produced in large numbers in the new military laboratory at Oak Ridge, in Tennessee, where they served to provide almost all the uranium-235 for the bomb which was later dropped onto Hiroshima on 6 August 1945, with tragic consequences.

---

[7] The concept of 'mass in motion' is a subtle one because the particle mass does not change during the acceleration. However, in a given magnetic field, the curvature of the trajectory increases *as if* the mass were to grow as the total energy: what is growing with the total energy is not the mass, but the inertia of the particle.

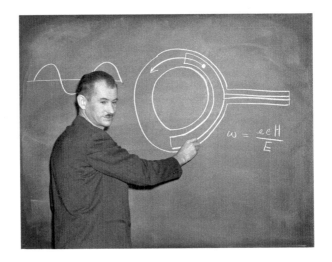

Fig. 1.14 Ed McMillan explains the principle of phase stability, which transformed the 184 in. cyclotron into a "synchrocyclotron" (Courtesy Lawrence Berkeley National Laboratory)

When, after the Second World War, Lawrence once again took up his original project, the problem due to the relativistic factor had found a solution: in 1944 and 1945 Vladimir Veksler in the Soviet Union and Edwin MacMillan in the United States had independently discovered the *principle of phase stability* (Fig. 1.14).

To understand the importance of this discovery, let us consider a bunch of particles, for example protons, which begin to circulate in a cyclotron. During the first turns the time that a bunch requires to travel a half-circumference is always the same because – with essentially constant total energy– the radius of the trajectory grows each turn in proportion to the increase in speed. However, when the kinetic energy exceeds about ten MeV, the total energy of the protons begins to increase and so too their inertia, i.e. their resistance to being accelerated; the bunch then requires more time to make a half turn and, at the subsequent transit between one electrode and the other, it does not arrive at the right moment to coincide with the accelerating voltage. This means that the dependence of transit time on the particle energy changes so that the arrival is no longer well synchronised with the kick imparted by the oscillating voltage.

This can be remedied by *increasing* the time between one oscillation of the applied voltage and the next, while the bunch of particles follows its spiral trajectory and the mass increases. This means decreasing the frequency of the radiofrequency source during the acceleration. However this would require a precision impossible to achieve in practice, if not for the phase stability principle. Let's see how this works.

At the gap between the cyclotron electrodes the voltage rises and falls at a fixed frequency. If a particle orbits *in phase* with this frequency, when it arrives at the gap it receives an accelerating kick and after a turn re-enters the space between the electrodes to receive the next kick at the same magnitude of the voltage. Suppose that another particle arrives at the gap *late*, when the voltage level is lower, then it will be accelerated less, and so follow an orbit of significantly smaller radius and require less time to complete a turn. When it arrives again at the gap it will therefore be *early*, and will find a higher voltage and be accelerated more. The consequence is that particles slightly out of phase oscillate turn after turn around those which are in phase, forming a compact and dynamically stable bunch.

This stability of the particles within a bunch is achieved when they cross the gap while the voltage is *decreasing*.[8] If this is the case, to keep the accelerated particles tightly bunched, it suffices that the oscillation period is increased, gradually and continuously, with the number of turns.

After the war Veksler and McMillan realised they had reached the same conclusions at the same time and, exchanging letters, became good friends. McMillan immediately recognized Veksler's scientific priority since his paper was published almost 1 year earlier in a Russian journal unknown in the United States. In 1963 they shared the Atom for Peace Award and McMillan commented on the simultaneity of the discovery with the following words:

> It seems to me that this is another case of a phenomenon that has occurred before in science – the nearly simultaneous appearance of an idea in several parts of the world, when the development of the science concerned has reached such a point that the idea is needed for its further progress. (David Jackson and Panofsky 1996)

The Synchrocyclotron Creates New Particles

As a consequence of the discovery of phase stability, Lawrence changed the original design of the electronic circuit, which applied the voltage to the electrodes, so that the oscillation period increased during the acceleration.

The 184 in. cyclotron was transformed in this way into the first 'synchrocyclotron', which began operation in 1946. Subsequently protons were accelerated up to 55 % of the speed of light, which corresponds to a kinetic energy of 200 MeV (Table 1.1). The energy of the accelerated protons was now, for the first time in an accelerator laboratory, large enough to produce new particles in the collisions with the nuclei of a target; this momentous step forward led the way to the discoveries of the next decade (Fig. 1.15).

When considering such a collision the first example which comes to mind is the impact of a car against a lamppost. In this case the damage is determined only by the kinetic energy. Since the masses of the car and the lamppost do not change, it is the kinetic energy of the car which disappears in the impact to produce the permanent deformation of the bodywork and the lamppost.

Instead, in the collisions between two particles, it often happens that other particles are created, as in the case of annihilation of an electron with a positron, whose energies are completely transformed into photons. In the phenomena produced by accelerated beams, the *total* energy, which is equal to the mass multiplied

---

[8] Similarly a surfer has to be on the decreasing part of an ocean wave to be transported towards the shore.

Fig. 1.15 The magnet of the 184 in. cyclotron (Courtesy Lawrence Berkeley National Laboratory)

by the relativistic factor, therefore counts more than the kinetic energy of the incident particle.

The energy of the protons produced by Lawrence's new synchrocyclotron was large enough to create, in the collision with the nuclei of a target, a great number of those new particles of which – only 1 year before – a few examples had been observed in photographic plates carried to high altitudes: *pions* (π).

Indeed in 1945 the photographic companies Ilford Ltd. and Kodak started the production of photographic plates having a sufficiently high concentration of silver bromide that the tracks left by all charged particles could easily be seen. Cecil Frank Powell, physics professor at Bristol and originator of the technique, described the events.

The young Italian physicist Giuseppe Occhialini *"immediately took a few small plates coated with the new emulsion and exposed them at the French observatory in the Pyrenees at the Pic du Midi, at an altitude of 3,000 m (where cosmic rays were little attenuated by the atmosphere). When they were recovered and developed in Bristol it was immediately apparent that a whole new world had been revealed. The track of a slow proton was so packed with developed grains that it appeared almost like a solid rod of silver, and the tiny volume of emulsion appeared under the microscope to be crowded with disintegrations produced by fast cosmic-ray particles with much greater energies than any which could be generated artificially at the time. It was as if, suddenly, we had broken into a walled orchard, where protected trees had flourished and all kinds of exotic fruits had ripened in great profusion."* (Powell 1972a)

In 1947 Cecil Powell, 'Beppo' Occhialini and César Lattes observed with their microscope events like those of Fig. 1.16.

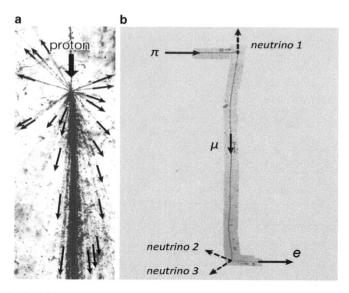

Fig. 1.16 On the *left* is a reproduction of a 'star' recorded in a nuclear emulsion. Some tracks, like the one enlarged on the *right*, show at their end two sequential decays: a pion produces a muon and the muon produces an electron. A muon lives for a microsecond and a pion a 100 times less ((**a**) Courtesy NASA; (**b**) Courtesy Indiana University – USA)

The dots, which form the tracks, are grains of emulsion which became black during development, because previously they were crossed by a charged particle.

In the photo on the left a very high-energy proton, originating from space, collides with an atomic nucleus. From the collision point a very large number of tracks emerge; some are left by protons from the nucleus, which has been split, and others from many charged particles created at the expense of the energy of the incident particle, among which are positively or negatively charged pions.

Following a few lighter tracks with the microscope, events similar to the one illustrated on the right are observed, which show a sequence of two decays. Today we know that the pion, slowed by collisions with the atoms of the emulsion, is stopped and then – in 20 billionths of a second – decays into a 'muon' (μ) and an electrically neutral particle, a 'neutrino' (v).

One notices that the neutrino, being uncharged, does not leave a trace in the emulsion; at the top of Fig. 1.16 it is represented by a dashed arrow. The muon, stopped in its turn, decays in two millionths of a second into an electron and two neutrinos.

One year before the beginning of operation of Lawrence's 184 in. synchrocyclotron, the pictures of Fig. 1.16 had caused enormous surprise, because at that time physicists knew from experiment only three types of subatomic particle: electrons, protons and neutrons. Cosmic rays were an extraordinary natural source of high-energy events, which were impossible to produce so far in the labs. However, the search for new particles using cosmic radiation as a source was a very difficult and time-consuming task, because the generated events were rare and their production uncontrollable.

All that changed in 1948 when a photographic emulsion was used as a target for a beam of particles generated by the 184 in. synchrocyclotron; on the developed plate hundreds of 'stars' were observed, from which emerged tracks that showed sequential decays.

Up to that time the beams produced by accelerators were used as substitutes for beams obtained from natural radioactive sources, to study Rutherford-like collisions with nuclei. In this way radioactive isotopes were also produced for employment in biology and medicine. With the advent of the Berkeley synchrocyclotron, and the production of proton beams of kinetic energy above 200 MeV, it became possible to produce and study phenomena involving the new subatomic particles. There was one single urgency: to experiment.

The advent of machines accelerating particles superseded the use of cosmic rays as a means for discovering new particles. This type of research came back into the laboratories and unfortunately lost the sense of adventure so well described by Cecil Powell:

> *Especially among some of the early contributors there was a marvellous sense of physical adventure. Their contributions seem to have something in common with what the Impressionists did for painting. Just as painters broke away from their studies and the limitations imposed by the academic style into the full light of the open air, so the cosmic-ray physicists escaped from their dark workshops, travelled the world, climbed mountains and made ascents in balloons.* (Powell 1972b)

An exciting era was closed and 1948 marks the birth of what is now called *high energy physics*, which from that time has been the main source of discoveries in laboratory experiments, using beams of artificially accelerated particles at ever higher energies.

Why 'Accelerators'?

To accelerate particles means to increase their speed, but the numbers of Table 1.1 show clearly such an increase in energy does not always correspond to a significant increase in their speed.

For example, when the total energy of a proton doubles, passing from 7.09 to 14.2 GeV, the increase in speed is only from 99 % to 99.7 % of the speed of light. Actually when the speed is close to that of light – the speed limit that no material body can reach – every further increase in total energy of the proton results in its greater resistance (inertia) to being deflected and accelerated.

Therefore, more precisely, the instruments we are speaking of in this book should not be called 'accelerators' but 'energizers' since they augment the energy of the particles rather than their speed. The term 'accelerator' is certainly more intuitive but is not correct, because it describes only the first and less significant part of the process. However its use is now so widespread to make it impossible to substitute.

Chapter 2
Small Accelerators Grow

Contents

© Springer International Publishing Switzerland 2015

35

U. Amaldi, *Particle Accelerators: From Big Bang Physics to Hadron Therapy*,
DOI 10.1007/978-3-319-08870-9_2

Daniele Bergesio, TERA Foundation

In 1960 the architecture of the first few CERN buildings was harmonious but rather spartan. A good forty years would pass before two well-designed buildings were constructed and called, in defiance of the disorderly numbering system typical of CERN, Building 40 and Building 39.

The physicists of the huge international collaborations working at the LHC occupy the first one, while the second is one of the hostels where some of the eight thousand physicists and engineers, who come from laboratories and universities all over the world to participate in the experiments carried out with CERN accelerators, stay for short periods. A large lawn and a road separate the two buildings from the main restaurant; and the area is traversed at all hours by many young researchers and, from time to time, also by some white-haired ones.

The square between Building 39 and 40 is dedicated to Edoardo Amaldi, who was one of the founding fathers of CERN and its Secretary-General between 1952 and 1954, in the years when the laboratory was created. As a young man Amaldi was part of the illustrious group of researchers gathered around Enrico Fermi in Rome, in the 1930s; they were famed as 'the boys of Via Panisperna', so named after the street where the Physics Institute was located. Other members of the group were Franco Rasetti, Emilio Segrè and Bruno Pontecorvo, whom we have either already met or shall encounter in later chapters.

Amaldi had an important role in the deliberations of May 1951, in which it was decided to propose to the CERN member states the construction of two accelerators, one intended for the present and the other for the future: a 600 MeV synchrocyclotron (which, after the entry of the Lawrence 184 inch machine into operation, was considered a sure success) and a much more challenging 10 GeV proton synchrotron, to whose story a section of this chapter is dedicated.

The Comeback of Linear Accelerators

After the invention of repeated acceleration, applied in the cyclotron of Ernest Lawrence, all the experts of the period were convinced that Wideröe's idea of the linear accelerator (Fig. 1.10) no longer had a future. All except one: Luis Alvarez, who was one of the most inventive scientists of the twentieth century. After the war he convinced himself that, at a sufficiently high energy, the cost of the circular accelerator would become prohibitive and therefore decided to use the new techniques of radiofrequency circuits to construct the first proton linear accelerator; after many highs and lows, in 1947 he succeeded to make it work.

Alvarez – who everyone called 'Louie' – was of Cuban-Spanish origin on his father's side and Irish from his mother. He had studied physics in Chicago after his father, a notable physician and medical researcher, left San Francisco in 1926 for the famous Mayo Clinic in Rochester, New York. Walter Alvarez had a great influence on the young Luis, including recommending him, as Luis recounts in his autobiography, *"to sit every few months in my reading chair for an entire evening, close my eyes, and try to think of new problems to solve. I took his advice very seriously and have been glad ever since that I did"* (Alvarez 1987).

During the Second World War Alvarez left Berkeley – where he had worked since 1936 – to contribute to the military development of radar at the MIT Radiation Laboratory in Boston. Among other things, there he invented and demonstrated the Ground-Controlled Approach system (GCA) which allowed landings of aircrafts at night and in conditions of low visibility, and the Vixen system which, mounted in a fighter aircraft, deceived the detection equipment installed on enemy submarines by creating the impression that the plane was receding while, instead, it was approaching. Having returned to Berkeley, he thought of using the thousands of radiofrequency components, of which the military had no further need, for the construction of his linear accelerator – the 'linac'.

Learning of this new idea, Lawrence, with his habitual enthusiasm, said: *"Alvarez has gotten the idea of putting obsolete radar equipment and incorporating it in a very interesting way in a Linear Accelerator which may make it possible to go to hundreds of millions of volts of all kinds of particles, not only electrons and protons but many ions as well. It is wonderful and in the Alvarez style – something that is out of the ordinary. It is bound to lead to very important physics in the future. The schemes of Alvarez and McMillan in my judgment is [sic] the outstanding event of their generation."* (Lawrence 1945)

The original idea of Alvarez had enthused Lawrence but did not work. The final one, which is still known as the 'Alvarez linac' is illustrated in Fig. 2.1a; in a copper tube – more than a metre wide and fifteen metres in length – are inserted large hollow cylinders of increasing length. As in the Wideröe linear accelerator, the proton bunches move at high speed inside the cylinders, receiving an accelerating kick at each gap between them and thus travelling increasing distances in each equal interval of time in which the electric field oscillates. However, the Alvarez accelerator tubes were not connected by cables to a source of oscillating high voltage, as in Fig. 1.10; instead a source of radio waves with 1.5 m wavelength created an electromagnetic field which oscillated 200,000 times a second along the entire length of the tube.

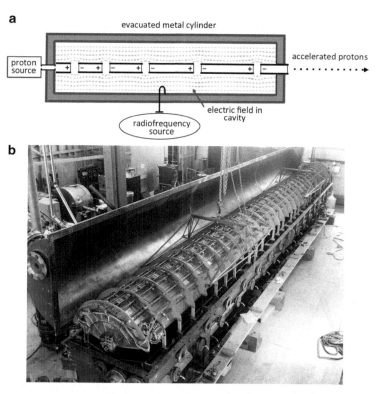

Fig. 2.1 (**a**) The principle of operation of an 'Alvarez'; the oscillating electric field produced in the cylinder by a radiofrequency (RF) source accelerates bunches of protons (electrically positive) when the left sides of the electrodes are positive. (**b**) The first proton linac under construction in the Berkeley Radiation Laboratory (Emilio Segrè Visual Archives/American Institute of Physics/Science Photo Library)

Fig. 2.2 (**a**) The Alvarez linac accelerated protons with a 200 MHz oscillating electromagnetic field (Courtesy Lawrence Berkeley National Laboratory). (**b**) The electron linac of Bill Hansen used an oscillating field with a frequency 15 times larger: 3,000 MHz or 3 GHz (William Webster Hansen Papers (SC0126), Courtesy Stanford University Archives)

This field produces an electric force, in the *gaps* between two successive tube sections, which accelerates the proton bunches – provided they traverse them at the right moment – and impels them to travel (with increased energy) along the length of the next section, within which they feel no force.

The dimensions of the first linac, which accelerated protons from 4 MeV up to 32 MeV, were impressive (Fig. 2.2a), and the number of particles accelerated each second was much larger than it was possible to extract from cyclotrons of the era, and for many applications this feature was extremely advantageous.

In the same years, just a short distance from Berkeley, at Stanford University, Bill Hansen – Luis Alvarez's teacher at MIT – built the first electron linear accelerator, which was based on a similar principle but, because an electron is two thousand times less massive than a proton, could use wavelengths about ten times smaller. Therefore the linac had a much smaller diameter, ten centimetres instead of about a metre. In the ever lively competition between Stanford and Berkeley, to poke fun at the picture of the Alvarez linac (Fig. 2.2a), Hansen had himself and three students photographed as they carried a section of his linac, which accelerated electrons up to 4.5 MeV (Fig. 2.2b).

Fig. 2.3 Seven Nobel laureates from the University of California, Berkeley, pose in front of the magnet from the 37 in. cyclotron (Courtesy Lawrence Berkeley National Laboratory, 1969)

The Hansen linac was immediately used for cancer therapy with X-rays and its story is told in Chap, 8, dedicated to medical uses of accelerators. However, before closing the interlude, let us return to Luis Alvarez, who was truly an eclectic genius (Fig. 2.3). Let three examples suffice.

First of all, as will be explained in the fourth chapter, in 1953 he launched the construction of a new type of detector, called 'bubble chamber', with which so many new particles were discovered that he earned the Nobel Prize. Secondly, in 1965 Alvarez proposed the use of cosmic ray detectors installed in the pyramid of Chephren in Egypt to make, essentially, a 'radiograph' of the upper part of the pyramid to seek any hidden chambers. The result – announced in 1969 – was negative but the method worked perfectly and was then applied to other pyramids.

Finally, Alvarez was seventy when his son Walter, a well-known geologist, showed him a sample of clay extracted in Gubbio (Italy), dating back 65 million years, the period in which the dinosaurs disappeared. Luis had the idea of analysing the sample to measure the concentration of iridium, a very rare element in the Earth's crust, but abundant in asteroids. Analysis using nuclear techniques revealed the presence of much iridium, not only in the Gubbio clay, but also in samples of rocks from the same period originating throughout the world. The conclusion was what is known today as the 'Alvarez hypothesis'; the iridium is of extraterrestrial origin and was spread across the globe 65 million years ago following the impact of an enormous meteorite. The residue from the impact darkened the skies for years, causing the extinction of dinosaurs and many other living species.

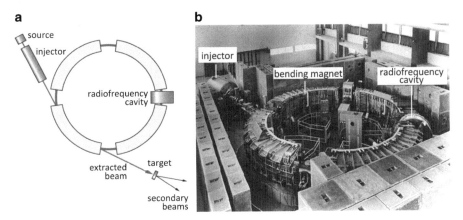

Fig. 2.4 Schematic diagram of a synchrotron and a photo of the electron synchrotron which began operation in 1959 at the Frascati laboratory of the Italian Institute for Nuclear Physics (INFN) ((**b**) Courtesy Physics Museum – Department Of Physics, University of Rome La Sapienza)

The First Synchrotrons

In a synchrocyclotron the spiral form of the orbits requires the deflecting magnetic field to be uniform over a large area, with the consequence that the weight and cost of the magnet which bends the trajectories increases greatly with the growth of the energy to be reached. In contrast, in place of a single large magnet, a *synchrotron* uses several smaller magnets placed around a circular 'doughnut', or hollow 'chamber' evacuated of air, in which the particles circulate.

As Fig. 2.4a shows, the principal components of this accelerating machine are the particle *source*; the *injector* – which gives the first acceleration to the particles and 'injects' them into the ring; the *bending magnets* – which steer the trajectories of particle bunches to keep them inside the ring; the *radiofrequency cavity* – made of two hollow electrodes which impart a small energy increment to the proton bunches, as they pass through them on each turn; the *extraction system* – which ejects the bunches of protons circulating in the ring when they have reached the desired energy, at the end of the acceleration cycle – and the vacuum chamber for the *extracted beam*.

In a synchrotron the weight of the deflecting magnets is much less than the weight of the single magnet of a cyclotron of equal energy and so, for the same cost and construction difficulty, much higher energy beams can be obtained. There are, however, two complications.

First of all, during the acceleration the magnetic field of the bending magnets must grow *in synchronism* with the increase in energy, so that the bunches of particles of steadily growing energy continue to follow, not a spiral orbit, but a *constant* circular path at the centre of the vacuum chamber. The growth of the magnetic field is obtained by increasing, during the acceleration, the current that circulates in the windings around the metal core, as in every electromagnet.

Fig. 2.5 At the International laboratory Joint Institute for Nuclear Research (JINR) in Dubna (Russian Federation) the Veksler Synchrophasotron operated until 2003 (Courtesy of JINR, Dubna)

In the second place, since the circulation time of the particles decreases with their increasing speed, the oscillation period of the voltage applied to the accelerating cavity must also diminish, *in synchronism*, with the energy increase, and therefore with the magnetic field; hence the name *synchrotron*. But, as in a synchrocyclotron, the synchronisation precision would not be sufficient were it not for the principle of phase stability, which guarantees that particles which are not exactly in time with the oscillation of the voltage on the cavity are also accelerated.

It is not surprising that the first synchrotron capable of beating the record energy of Lawrence's 184 in. synchrocyclotron should have been built by one of the discoverers of this principle. Between 1945 and 1948, under the direction of Ed McMillan – who had invented the names 'synchrocyclotron' and 'synchrotron' – a 340 MeV electron synchrotron was constructed at Berkeley. With it a particle never before observed in cosmic rays was discovered: the *neutral pion*.

In 1949 the 250 MeV electron synchrotron designed by the other discoverer of the phase stability principle, Vladimir Veksler, began operation in the Soviet Union. In the same year he planned a proton synchrotron with a 200 m circumference, the 'Synchrophasotron', which in 1957 reached 10 GeV energy (Fig. 2.5).

In Europe the first electron synchrotron was demonstrated in 1946 by F.G. Goward and D.E. Barnes at Woolwich Arsenal Research Laboratory in England, and the first proton synchrotron, constructed at the University of Birmingham by M. Oliphant and his team, accelerated protons up to 1 GeV of energy in 1953.

For many years the race towards ever higher energies was dominated by the United States. In particular, in 1946 nine large universities (Columbia, Cornell, Harvard, Johns Hopkins, Massachusetts Institute of Technology, Princeton,

University of Pennsylvania, University of Rochester and Yale) created a non-profit organisation with the role of fundamental research into nuclear sciences, encompassing physics, engineering, chemistry and biology, and the construction (on the site of the former Camp Upton military base) of large facilities – specifically an accelerator and a nuclear reactor – that no single university could afford to develop alone. The laboratory took the name of Brookhaven National Laboratory (BNL) and its creation has been described by the American Nobel laureate Norman Ramsey:

> *The idea that grew into Associated Universities Inc. (AUI) and Brookhaven National Laboratory (BNL) arose in discussions between Isidor Rabi and myself at Columbia University during the period from October to December of 1945, shortly after Rabi returned to Columbia from the MIT Radiation Laboratory and I returned from Los Alamos. I wish I could claim that these discussions originated in a flash of genius and in a vision of AUI and Brookhaven as the important institutions they are today. Instead, I must admit that the idea grew from a mood of discouragement, jealousy and frustration. In particular, Rabi and I both felt that physics at Columbia University was coming out in this period with little scientific benefit in return. In particular, the earliest United States work on fission and nuclear reactors was that done at Columbia by Fermi, Szilard, Zinn, Dunning, and others. However, during the course of the war this activity had been transferred to other locations. As a result many universities emerged from the war with strong nearby nuclear science research laboratories, whereas Columbia did not. (Ramsey 1966)*

The BNL synchrotron – which became operational in 1952 and later accelerated protons up to 3.3 GeV – was dubbed the 'Cosmotron' to indicate that the machine would deliver beams of particles having energies as large as those of the cosmic rays which, at that time, were the means of discovery of new particles. The protons were kept in a circular orbit by 2,000 t of deflecting magnets, less than half the weight of the magnet of the 184 in. synchrocyclotron, which reached only 0.2 GeV.

Only 1 year after, at Berkeley, the 'Bevatron' accelerated protons up to 6.3 GeV, the energy chosen because, in the collision with target nucleons, the protons would liberate the 2 GeV which are necessary to create a proton-antiproton pair.

The magazine 'Popular Science' described the technical marvels of the Bevatron in an article entitled "A 10,000-ton Cracker for Invisible Nuts":

> *In a hollow in a California hillside a large but unpretentious building covers one of the biggest and strangest of all machines. It is 135 feet across, cost 9,500,000 dollars and contains more than 9,500 tons of iron, 225 miles of wire and 2,400 vacuum tubes. Its 31 vacuum pumps evacuate the equivalent of a seven-room house. What it makes cannot be seen or felt, let alone be sold. This is the Bevatron, just completed, the most powerful atom-smasher yet built. (Huff 1954)*

As expected, the existence of antiprotons was proven in the collisions of 6.3 GeV protons with the nuclei of a solid target by a group of physicists led by the Italian Emilio Segrè (who had trained under Enrico Fermi in the 1930s as one of the 'boys of Via Panisperna') and made up of Owen Chamberlain, Clyde Wiegand and Tom Ypsilantis, a brilliant young doctoral student of Greek origin born in the United States. The discovery was rewarded by a Nobel Prize in 1959, shared by Chamberlain and Segrè (Fig. 2.3).

In Italy, things moved much more slowly because after the war the country was in ruins and funding was not available for science until the years of the post-war economic boom. In the reconstruction surge of the 1950s, physicists from Rome, Florence, Turin and Padua, who were mostly occupied with cosmic ray research, founded the Italian National Institute of Nuclear Physics (INFN) in 1951, whose first president was Gilberto Bernardini and whose second was Edoardo Amaldi.

At the beginning of 1959 a 1 GeV electron synchrotron started operating in the new Laboratory built by INFN in Frascati, close to Rome (Fig. 2.4b). It was a late development and the energy was not particularly high for the times, but I mention it here because this machine was instrumental in initiating my long-term involvement with particle accelerators.

The Invention of Strong Focusing

When the Cosmotron began operation at Brookhaven National Laboratory it was calculated that to reach an energy ten times larger (30 GeV) it would be necessary to employ a hundred times more iron (200,000 t).

During the acceleration, trajectories of particles, which circulate for a million turns inside the vacuum chamber of a synchrotron, do not remain constantly at the centre of the bending magnets, but oscillate vertically and radially about the central orbit. In order that the protons were not lost by collisions with the walls of the chamber, the cross-section of the Cosmotron ring was very large, around 20 cm vertically and 60 cm horizontally. To produce the necessary magnetic field on a ring of these dimensions, 2,000 t of iron were required. At higher energies the oscillation amplitudes increase further and so it becomes necessary to increase the chamber dimensions, and consequently, the amount of iron used.

In the summer of 1952 a delegation of accelerator experts from the newly created CERN (*Conseil Européen pour la Recherche Nucléaire*) laboratory went to Brookhaven to visit the Cosmotron and discuss with their American colleagues the properties of the European accelerator under construction. While preparing for this visit, Stanley Livingston (a former student of Lawrence), Ernest Courant and Hartland Snyder from Brookhaven had a brilliant idea for reducing the oscillation amplitudes and thus the cost of the magnets. The visitors brought this new concept back to Europe that, as I shall explain later, had a great impact on the CERN programme and on the building of the first European synchrotron. Meanwhile, the three authors published a scientific paper in 'Physical Review', the most important US physics journal, entitled "The Strong Focusing Synchrotron – A New High Energy Accelerator". In essence, the three suggested the use of a *transverse* magnetic field in addition to the *vertical* magnetic field which maintained the particles in their circular orbit.

To explain how *strong focusing* works it is helpful to refer to the focusing of a beam of particles, which, in the absence of a vertical magnetic field, propagates

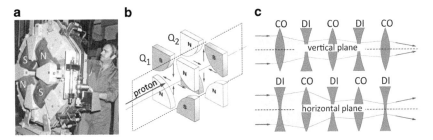

Fig. 2.6 The currents circulate in the copper windings of the 'quadrupole' in such a way that it has two South poles and two North poles facing each other, which produce magnetic fields – at the centre of the quadrupole – transverse to the direction of motion of the particles (Courtesy CERN)

horizontally in a straight line. The photograph of Fig. 2.6 shows a 'quadrupole', a magnet that has two North (N) poles and two South (S) poles.

Two quadrupoles on the same central axis produce transverse magnetic fields which run from South to North poles. These fields give rise to forces in quadrupole Q1, as shown by the arrows in the drawing of Fig. 2.6, which are directed towards the axis in the vertical plane and away from the axis in the horizontal plane.

The drawing on the right shows how the vertical and horizontal dimensions of a parallel particle beam are reduced under the effect of a series of quadrupoles which alternate between converging (CO) and diverging (DI), and so on, in the vertical plane. The focusing *also* takes place in the horizontal plane, where the first quadrupole is diverging (DI).

It may seem surprising that an alternating sequence of converging and diverging 'magnetic lenses' always has an overall convergent effect. This is the brilliance of the idea, which Courant, Livingston and Snyder confirmed with the many equations of their Physical Review article. A similar focusing effect can be obtained with beams of light by using an alternating series of optical lenses.

To illustrate the advantages of this new approach to particle acceleration, Fig. 2.7 compares a modern strong focusing synchrotron with a normal synchrotron, henceforth referred to as 'weak focusing'.

In the strong focusing synchrotron the more numerous bending magnets alternate with quadrupoles to keep the proton bunches well focused during each acceleration cycle. The protons are not lost, despite the vacuum chamber of the ring being much narrower than in a weak focusing synchrotron, typically 5 cm vertically and 15 cm horizontally (Fig. 2.8).

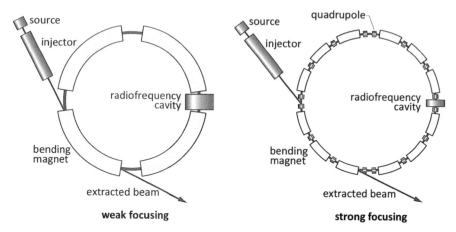

Fig. 2.7 Schematic comparison between a weak focusing synchrotron (like the one in Fig. 2.4) and a strong focusing machine which achieves the same energy

Fig. 2.8 Snyder, Livingston and Courant (from *right* to *left*) are pictured in front of the mock-up of a strong focusing bending magnet (From Panofsky) (Courtesy of Brookhaven National Laboratory)

Phase stability and strong focusing are the ideas on which all high-energy accelerators are now based. They are of very different degrees of sophistication, as remarked by McMillan (Fig. 2.3).

> *In the case of phase stability, it is easy to approach this in an intuitive, elementary way. When I first started telling people at Los Alamos about it, I found that I hardly needed to finish the explanation before they understood, and at least one said to me that he felt stupid in not having thought of it himself. But strong focusing is not the kind of thing that one thinks of in an elementary way.*
>
> *The team that did this in the United States was Stan Livingston, Ernie Courant and Hartland Snyder. But there was an independent inventor. He was working entirely isolated from contact with anybody, in Greece, and was somewhat earlier than these three. That was Nick Christofilos, one of the real geniuses of our time.* (McMillan 1973)

Interwoven Histories

The article by Courant, Livingston and Snyder was enthusiastically welcomed by the physics community, which was eager to exceed the 10 GeV barrier without recourse to the 40,000 t of iron needed by the Synchrophasotron. Everyone was convinced that the three inventors would soon share a Nobel Prize, but that did not happen because the idea had already been patented 2 years earlier by the Greek engineer Nicholas Christofilos.

Born in the USA in 1916, Christofilos had returned to Athens with his parents at the age of seven. He graduated from the National Technical University of Athens in electrical and mechanical engineering, later working as a lift engineer and repairing German lorries during the Nazi occupation. In his free time, he studied the physics textbooks he was able to find and, completely self-taught on the subject of particle accelerators, at the age of thirty he filed a patent, which essentially applied the phase stability principle to devise independently the idea of the synchrotron. While he built up his own lift maintenance company he learned, reading scientific magazines, that the synchrotron had already been invented. Instead of discouraging him, it drove him to search for a means of focusing protons accelerated in synchrotrons.

In 1948 Nick Christofilos wrote a long letter to scientists at Berkeley in which he described his own concept, which proved to be flawed. Having received a detailed answer explaining his mistakes, he continued to think about the problem, and while doing so discovered strong focusing. Therefore, the next year, he wrote another letter to Berkeley describing his latest ideas; this time, however, no one paid attention to the contents, even though correct, because the mathematics was complicated. Not receiving an answer, in 1950 Christofilos filed a second patent.

Three years later, while visiting the United States, he read the Physical Review article of Courant, Livingston and Snyder in the Brooklyn Public Library and became convinced his idea had been stolen. He rushed to Brookhaven where he learned that the Americans had arrived at the same conclusion without knowing of his patent. He was immediately offered a position, obtained recognition of his priority and $10,000 from the Atomic Energy Commission (*AEC*) for exploitation of the patent in the new Brookhaven synchrotron.

In Greece he had developed another idea in a field that at that time was just emerging: a new type of reactor that he called 'Astron' for the production of energy from nuclear fusion processes. This time the AEC listened attentively to him, even more because Astron seemed to solve problems that were being encountered in the implementation of two American-conceived reactors, the 'stellarator' and the 'magnetic mirror'. Astron was built in the Livermore Laboratories, in California, and to realise it Nick Christofilos invented a new type of linear induction accelerator, which is still in use today.

In Livermore he also worked on national defence, proposing the creation of an artificial belt of electrons around the earth produced by a nuclear detonation.

TRIUMPH IN SPACE FOR A 'CRAZY GREEK'

Theory of Boston-born maverick scientist led to sensational Project Argus

Fig. 2.9 This picture of Nicholas Christofilos explaining the Argus project appeared in the March 1959 issue of *Life* magazine (Trombley 1959)

In 1958, an exploratory experiment, called Argus, was carried out by exploding three small nuclear bombs above the South Atlantic Ocean (Fig. 2.9).

Despite all his efforts his pet project, Astron, never worked. He directed it with great determination, working day and night, defending it against detractors, arguing vigorously and drinking heavily, until a massive heart attack killed him in 1972 at the age of only 55. Shortly before he had met the head of the AEC, who had informed him of his project's cancellation. The life of this self-taught genius, often overlooked in the history of accelerator designers, has many aspects of a Greek tragedy.

How My Passion for Accelerators Started

The Courant, Livingston and Snyder article gave rise to the sequence of events just described and had many other consequences. The most important concerned the construction of the first CERN synchrotron, as we will see later. It also had an influence on me personally when, in 1957, Mario Ageno, head of the physics laboratory in the Italian National Health Institute (ISS), asked me – who had just arrived – to study its details, with the objective of designing and constructing a focusing system for the injector of the Frascati synchrotron, then under construction (Fig. 2.4).

It might seem odd that a group of researchers in a medical institute would be engaged in the construction of an accelerator. To understand the reasons, it is necessary to think back to the middle of the 1930s, when Enrico Fermi and the 'boys of via Panisperna' in Rome discovered that slow neutrons were more efficient than fast ones for the production of radioactive materials. Fermi, Franco Rasetti and Edoardo Amaldi therefore designed a 1 MeV energy Cockcroft-Walton accelerator to provide a very intense source of slow neutrons, and constructed it not at the University of Rome La Sapienza, but in the National Health Institute, which obtained the required funds from the Ministry of Health; indeed the newly abundant radioisotopes would be used for irradiating tumours. These developments were following the track opened by Ernest Lawrence in the States with the second generation of his cyclotrons.

In 1938 Fermi, whose wife was Jewish, left Italy for good in reaction to the racial laws to go to Columbia University in New York, and Edoardo Amaldi completed and operated, with the young Mario Ageno, the first Italian accelerator. Fifteen years later, when the Frascati synchrotron was launched, it was therefore natural that Ageno should lead a small group of researchers in the construction of an accelerator, similar to the one still then in operation at the ISS, which had the job of 'injecting' electrons into the, thousand times more powerful, second Italian accelerator.

In 1957, I had graduated a few months earlier and had decided to work at the ISS, because I did not want to participate in a research group in which my father Edoardo Amaldi, then already well known, was active. It was an excellent opportunity, given that Mario Ageno was an outstanding scientific personality from whom I learned a great deal.

I can still remember how much effort it took me to understand the strong focusing principle embodied in the equations of that famous article. Muddling through as best I could, in a couple of months I succeeded in converting some of the formulae into a mechanical design for two quadrupoles about ten centimetres long. Today I can say with satisfaction that those were the first quadrupoles built in Italy and that some examples have been working for many decades. Certainly their construction contributed to my deep interest in particle accelerators and these instruments have continued to fascinate me both for their use in fundamental physics and for their medical applications.

In the field of fundamental physics, I have had the good fortune to participate in, and sometimes to design and guide, experiments in nuclear and sub-nuclear physics carried out at several major accelerators: at Frascati the electron-synchrotron (Fig. 2.4); at CERN the Proton Synchrotron (PS), the Intersecting Storage Rings (ISR), the Super Proton Synchrotron (SPS) and the Large Electron Positron collider (LEP).

As far as medical applications are concerned, in the last 20 years I have worked on proton and carbon ion accelerators dedicated to the treatment of solid tumours; one of the accelerators, a synchrotron, which my research group and I designed, is now in operation at the National Centre for Hadron Cancer Therapy (CNAO), in Pavia, where patients have been treated since September 2011. The second accelerator we have been working at is a novel proton linac; the company A.D.A.M., spin-off of CERN, started its construction in 2014.

Accelerators for fundamental physics and medical treatments are based on the same principles of operation – in particular phase stability and strong focusing – and on the same techniques, but the objectives are different.

The first type of application targets the construction of what I call *beautiful physics*, driven by the innate need of comprehending and explaining the world around us, which provides the incomparable pleasure of understanding, and sometimes actually discovering, new natural phenomena.

The second falls instead into the category of what I call *useful physics*, meaning applied physics in general but, for me, specifically those applications which permit new methods of diagnosis and treatment of otherwise incurable illnesses. I am happy to have had the opportunity to dedicate my professional life to these two fascinating and interwoven branches of physics.

The Birth of CERN

In the years between 1945 and 1950, the idea of an international laboratory dedicated to physics using accelerators sprang into life more or less independently in several European circles. Many scientists and some politicians were worried by the conspicuous disparity which was developing between research in the United States and that in Europe. Moreover the flight of intellectuals resulting from the difficult pre-war and post-war conditions was rendering the recovery of top-level scientific research in Europe even more challenging.

The first seed was sown in 1945 by the American physicist Robert Oppenheimer, who had directed the Manhattan Project at Los Alamos, and who, in the course of work in the UN Commission for the Control of Nuclear Energy, formed a friendship with the French diplomat François de Rose, convincing him of the necessity for the European countries to work together towards the construction of the large and expensive instruments by then necessary to match research in the USA. This message was passed on to French physicists who participated in the work of the UN Commission – in particular to Pierre Auger, Francis Perrin and Lew Kowarski – and in October 1946 the French delegation submitted to the UN Economic and Social Council a proposal to create the 'United Nations Research Laboratories', dedicated especially to the peaceful development of nuclear energy.

In December 1949, the writer and philosopher Denis de Rougemont organised a European Conference of Culture at Lausanne. On this occasion, Raoul Dautry – Director General of the French Atomic Energy Commissariat– read out a letter from the great French theoretical physicist Louis de Broglie, the discoverer of the wave nature of the electron, who was unable to attend the conference in person:

> *At a time in which we speak of the union of the peoples of Europe, the problem arises of extending this new international unity by the creation of the laboratory or an institution where it would be possible to carry out scientific work, in some manner beyond the scope of the individual participating nations. As a result of the cooperation of a large number of European countries, this body could be endowed with resources beyond the level available to national laboratories.* (de Broglie 1949)

Among the objectives of the international laboratory, Dautry conjectured, on one side, research in astrophysics through the construction of powerful telescopes; on the other, research in the field of energy, given the growing concern about the steady reduction of natural resources, especially coal. The concluding motion therefore proposed the future Institute of Natural Science should be directed towards 'applications useful for everyday life'. This, however, was not the general tendency of the initiatives which were developing in other European circles.

In Rome, at the end of the 1940s, Edoardo Amaldi had begun to discuss with some of his colleagues the need to create a European laboratory of fundamental physics, equipped with the accelerators which could compete with the best large American centres, in particular with Brookhaven National Laboratory, which was completing its Cosmotron.

As a high school student I had often heard my parents talk about these subjects and I took part, during our frugal daily meals, in animated discussions with notable European physicists. Among them, I remember the Frenchman Pierre Auger and the Englishman Cecil Powell, who had been awarded the Nobel Prize in 1950 for the discovery of the charged pion using nuclear emulsion techniques (Fig. 1.16).

In June 1950 the Nobel Laureate Isidor Rabi – one of the founding fathers of Brookhaven National Laboratory – participated as United States delegate in the General Assembly of UNESCO (United Nations Educational, Scientific and Cultural Organization) held in Florence. Having previously discussed the subject with the people mentioned above and having obtained the authorization of the American government, Rabi proposed that UNESCO should "*assist and encourage the formation of regional research laboratories in order to increase international scientific collaboration*" (UNESCO 1950). In the discussions in Florence there had been no mention of particle accelerators but three days later, during a press conference, Rabi spoke explicitly of this possibility.

To discuss Rabi's proposal, in December Denis de Rougemont – who had become Director of the newly created European Cultural Centre – assembled in Geneva with Auger's help a group of Belgian, French, Italian, Norwegian and Swiss physicists. The final document recommended "*the creation of an international laboratory centre based on the construction of an accelerator capable of producing particles of an energy superior to that foreseen for any other accelerator already under construction*" (Centre Européen de la culture 1950). The cost estimate (of 20–25 million dollars) was taken from the paper prepared by the Italian Bruno Ferretti, professor of theoretical physics in Rome.

This proposal was taken forward at the end of 1950 by Pierre Auger who – at the time – directed the Natural Science division of UNESCO. He could act because, immediately after the Florence assembly, about $10,000 were donated to UNESCO by Italy, France and Belgium for the organization of the first meetings of a group of consultants from eight European countries.

Auger was a master of experimental physics; among other things he had discovered that high-energy cosmic rays produced cascades of particles in the atmosphere, which had been named 'extensive air showers'. It was impossible not to be impressed by Auger; he was tall, with a beard making him resemble a cartoon scientist, and came from a cultivated background, writing poetry in his spare time. He made me the present of a book, when I was a teenager passing a summer in his family holiday home in Brittany. The house was located in a village called, by physicists who knew of it, 'Sorbonne on Sea' ('Sorbonne plage') because it was frequented by the families of eminent Parisian professors, among whom – in the 1920s and 1930s – was Marie Curie.

In May 1951 in Paris, the seat of UNESCO, the 'Board of Consultants' met for the first time and two goals were established: the very ambitious project to build a proton synchrotron second to none in the world and, in addition, the construction of a 'standard' machine – a synchrocyclotron – which would allow an early start to experimentation. The government delegates met twice more under the auspices of UNESCO, which invited all its European members, including the countries of Eastern Europe; these, however, did not show up with the exception of Yugoslavia. Thus twelve countries from Western Europe were represented at the two conferences held in Paris at the end of 1951 and in Geneva at the beginning of 1952.

Auger and Amaldi wanted from the outset to create a truly international laboratory dedicated to particle accelerators, preferably in a neutral and geographically central location like Switzerland (Fig. 2.10). The physicists from northern Europe, gathered around the great Niels Bohr, preferred instead the idea of a decentralised organisation, which would exploit infrastructure and accelerators in already existing laboratories. There were discussions and exchanges of very heated letters but, in the end, the idea of a centralised structure prevailed, thanks to the speed with which the southern European physicists moved.

Eventually the agreement to create the 'Conseil Européen pour la Recherche Nucléaire' (CERN) was signed in Geneva in February 1952 and nominations made. Edoardo Amaldi became Secretary General of the provisional organisation; the Dutch Cornelis Jan Bakker was nominated director of the Synchrocyclotron group, the Norwegian Odd Dahl director of the Proton-Synchrotron group, the Frenchman Lew Kowarski director of the Laboratory group and Niels Bohr as director of the Theoretical group. On that occasion a telegram was sent to Isidor Rabi:

> We have just signed the Agreement which constitutes the official birth of the project you fathered at Florence. Mother and child are doing well, and the Doctors send you their greetings. (CERN Archives 1952)

The CERN Convention was established in July 1953 and signed by the 12 founding Member States: Belgium, Denmark, France, the Federal Republic of Germany, Greece, Italy, the Netherlands, Norway, Sweden, Switzerland, the United Kingdom and Yugoslavia. The ratification process by the Parliaments was lengthy– at least in the mind of the promoters – so that only on 29 September 1954 was the required number of ratifications reached, and the European Organization for Nuclear Research then came officially into being. However the old acronym CERN was kept, even if it is now read 'European Laboratory for Particle Physics'.

Sixty years on, around ten thousand physicists and engineers, originating from almost every country in Europe and the world, work at CERN – the largest physics laboratory in the world – demonstrating that the original guidelines so determinedly

Fig. 2.10 Pierre Auger (on the *left*) with Edoardo Amaldi at the meeting of the CERN Council held in Paris in 1953 (Courtesy Edoardo Amaldi Archive – Department Of Physics, University of Rome La Sapienza)

upheld by Amaldi and Auger have proven successful. Carlo Rubbia – Nobel Prize for physics in 1984 and Director General of CERN from 1989 to 1994 – has written in a biography of Edoardo Amaldi, published by the Royal Society:

> *Amaldi was inspired by two clear principles. First he was convinced that science should not be pursued for military purposes. Secondly, Amaldi was a dedicated European. He realised very early that no single European nation could hope to hold its own scientifically and technologically. To compete they had to collaborate. The [CERN] laboratory has become the greatest physics centre in the world, where more than half the particle physicists on our planet work. The dream of Amaldi to re-establish a centre of excellence and to halt the 'brain drain' to the United States has been completely realised.* (Rubbia 1991)

It was actually in 1989, when the new LEP accelerator began operation, that the flow of scientists reversed, as shown in Fig. 2.11. Since then the number of American physicists who work in Europe has been larger than the number of European physicists who work in the USA.

Today the number of CERN member states is twenty-one, nine more than at the beginning. At the end of 2012 Israel and Serbia were appointed Associate Members in the pre-stage to membership; in 2014 Israel joined as full member. In addition the United States, Japan, the Russian Federation, India and Turkey participate with the status of 'observer states' in the meetings of CERN Council and have thousands of researchers performing experiments at CERN. All these countries contributed – with many others [1] – to the construction of the latest accelerator, the *Large Hadron Collider* or LHC, and its detectors.

---

[1] Non-member states with co-operation agreements with CERN include Algeria, Argentina, Armenia, Australia, Azerbaijan, Belarus, Bolivia, Brazil, Canada, Chile, China, Colombia, Croatia, Ecuador, Egypt, Estonia, Former Yugoslav Republic of Macedonia (FYROM), Georgia, Iceland, Iran, Jordan, Korea, Lithuania, Malta, Mexico, Montenegro, Morocco, New Zealand, Pakistan, Peru, Saudi Arabia, Slovenia, South Africa, Ukraine, United Arab Emirates and Vietnam.

Fig. 2.11 The number of particle physicists originating in the USA who work in Europe between the second half of the 1970s and the beginning of the 1990s, compared to the number of Europeans who work in the United States. In 2012 about 1,000 American physicists and engineers were carrying out experiments at the CERN accelerators, mainly LHC

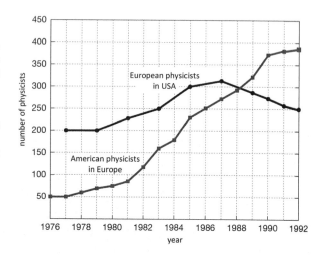

Choice of the Large Accelerator

In August 1952, awaiting the ratification of the international treaty by the parliaments of the twelve member states, which would formally approve the new organisation, a delegation of experts visited the then unfinished Brookhaven Cosmotron and learned about the recent invention of strong focusing. The visitors were prestigious: the Norwegian Rolf Wideröe, inventor of the linear accelerator which had inspired the young Lawrence, the Englishman Frank Goward, who had built the first European electron-synchrotron, and another Norwegian: Odd Dahl, leader of the delegation. Odd Dahl was a remarkable character, with a strong physique and adventurous spirit (Fig. 2.12). In 1922, at the age of 24, he decided to join the expedition to the North Pole organised by the great explorer Roald Amundsen, as the pilot of a small plane. On board the icebreaker *Maud*, aircraft and crew let themselves drift at the edge of the pack ice, so as to approach as close as possible to the Pole.

With the boat trapped by ice, the plane permitted exploration of the surrounding area; these were among the first polar flights to take off from a ship. The event which signified the start of Dahl's scientific career occurred a few months after the departure of the expedition; during a landing the plane broke up, forcing Dahl to remain on board the ship for more than 2 years as a prisoner of the ice, together with other members of the expedition. Despite lacking any scientific training, he used this time to study physics, constructing very sophisticated instruments and undertaking geophysical observations. He recounted that these were very productive years, completely dedicated to study and work, even if – being the youngest member of the team – he did not have the right to begin a conversation with his fellow adventurers.

On his return, he went to Washington to work on one of the first electrostatic accelerators and became a renowned expert at international level. Returning to

Fig. 2.12 In the
Smithsonian Institution
Archives one can find this
picture of Odd and his wife
Anne "Vesse" Dahl
(Courtesy of Smithsonian
Institution Archives)

Norway in 1936, he built three particle accelerators and later, after a few years
working in CERN, he led the construction of the first nuclear reactor built by a small
country.

Kjell Johnsen, Dahl's most brilliant pupil, described the impact of the invention
of strong focusing on the construction of the CERN synchrotron in this way:

> Dahl at once saw the implications and convinced his group that this was the way to go. All
> effort was immediately switched. That autumn, CERN's Council was also convinced, and
> one of the most important decisions in CERN history was made. Intuition governed the
> choice more than knowledge. It would have been much easier (as other laboratories did and
> later regretted) to have played safe. Had CERN gone for a 10–15 GeV scaled-up
> Cosmotron, its future would have been very different. It was also a very unselfish decision
> for Dahl, because the whole nature of the Proton Synchrotron Group work changed.
> Instead of being essentially an engineering group scaling up an existing machine based
> on well-established principles, it became a physics group studying the theory of acceler-
> ators, only later returning to engineering design. To lead this demanded full-time commit-
> ment, which Dahl could not give, and he returned to Norway. (Johnsen 1994)

In October 1952, during the third session of the CERN Council meeting held in
Amsterdam, the Council decided to construct – without increasing the cost – a
25 GeV synchrotron based on the new ideas, instead of the 10 GeV weak focusing
synchrotron which was foreseen before the Brookhaven visit. A similar decision

had been taken at Brookhaven for an increase to 30 GeV energy of the original 3 GeV Cosmotron and the construction work for the two synchrotrons – European and American – started at the same time. It is interesting to note that during the same Council meeting Geneva was chosen as site of the European laboratory.

The CERN Proton Synchrotron (PS), which has a circumference of more than 600 m, started operation in 1959, while the AGS ('Alternating Gradient Synchro-tron') in Brookhaven, of about 800 m length, began to accelerate particles several months later.

Over more than 50 years of operation, the Proton Synchrotron, the historic first CERN machine, has accelerated all types of particle: electrons and positrons, protons and antiprotons, as well as heavy nuclei such as lead.

John Adams and the PS

The construction of the Proton Synchrotron was led by John Adams, a self-made man who – with a war invalid father – could not go to university and in 1939 had obtained a modest Higher National Certificate, at the age of nineteen, at a London technical college. After having worked during the war on radar development, in 1945 he was employed by the UK Atomic Energy Research Establishment (AERE) in Harwell, gaining the responsibility of lead engineer in charge of construction of the first large European accelerator, a 175 MeV synchrocyclotron which began operation in 1949. His success in this difficult challenge drew him to the attention of the Harwell director, Sir John Cockcroft, the Nobel Laureate who had built the first electrostatic accelerator with Walton.

Adams was recruited to CERN at the age of 33, following a meeting organised by Cockcroft with Amaldi, who was then 48 years old. Amaldi described the events of that day:

> I met John Adams for the first time on 11 December 1952 in London at the Savile Club, where both of us were invited to lunch by John Cockcroft. On the telephone I had expressed to Sir John the desire to exploit my trip to London to meet some young British physicists and engineers who could be interested in participating in the construction of the European Laboratory. At lunch I was immediately impressed by the competence of John Adams in accelerators, his open mind on a variety of scientific and technical subjects, and his interest in the problem of creating a new European laboratory.
>
> In the early afternoon I was received, together with Ben Lockspeiser - Secretary of the Department of Scientific and Industrial Research (DSIR) - and Sir John, by Lord Cherwell, scientific adviser to the then Prime Minister, Winston Churchill. As soon as I was intro-duced to him in his office, he said that the European laboratory was to be one more of the many international bodies consuming money and producing a lot of papers of no practical use. I was annoyed, and I answered rather sharply that it was a great pity that the United Kingdom was not ready to join such a venture, which, without doubt, was destined for full success, and I went on by explaining the reason for my convictions. When we left the Ministry of Defence I was unhappy about my lack of self-control, but Sir John and Sir Ben were rather satisfied and tried to cheer me up.

Shortly after 4 p.m. I left London for Harwell. The conversation during the three hour drive confirmed my first impression of John Adams; he was remarkable by any standard, and he was ready, incredibly ready, to come to work for CERN. I was also impressed by the other young people I met at Harwell. Contrary to the impression that I had got from Lord Cherwell early in the afternoon, they were not at all insularly minded. (Amaldi 1986)

After a few weeks Sir Ben sent a letter requesting observer status for the UK in the provisional organisation and the DSIR (Department of Scientific and Industrial Research) began to contribute 'gifts', as they were called, which precisely corresponded to the fraction of the required investment, calculated according to the same rules applying to the other eleven countries.

In September 1953 John Adams became a CERN staff member and was made responsible for construction of the PS when, in March 1954, Frank Goward died suddenly after succeeding Odd Dahl in leading the acceleration group.

Before arriving in CERN Adams had already made a great contribution to the planning of the PS when, in winter 1952, it was discovered that the application of strong focusing, adopted in the first design, introduced such instability that the proton beams would have been lost during the acceleration. The magnet design was changed, such that the focusing was made much less strong, and the weight of the magnets increased from 800 to 3400 tons. In this, and in many other difficult decisions that became necessary, Adams demonstrated all his qualities as an engineer and a manager, maintaining his calm even in the most stressful moments and identifying – with the contributions of all – the most appropriate engineering solution.

The CERN accelerator experts have always been of very high standard, but I do not think I exaggerate in claiming that the success of all the accelerators built by CERN, and especially the LHC, have their roots in the quality of the work carried out under the guidance of John Adams, during the 5 years of the PS construction (Fig. 2.13).

The PS began operation on 24 November 1959 while the AGS (Alternating Gradient Synchrotron), constructed in parallel at Brookhaven, accelerated a beam of protons to 33 GeV 6 months later. Despite the sophistication of the theory of strong focusing synchrotrons, at that time many aspects of the behaviour of proton bunches during the acceleration were still very obscure, so that in the first Quarterly PS Report one can read a sentence, possibly written by Adams (Fig. 2.14): *"The situation in December 1959 was that the synchrotron had worked successfully to its design energy, and already beyond its design current, but with its builders and operators in a state of almost complete ignorance on all the details of what was happening at all stages of the acceleration process."* (CERN PS 1960)

At this point I want to stress that, the PS would never have been built so efficiently and in such a short time without the active help and presence at CERN of Brookhaven specialists; in the relations between the two laboratories the collaboration was more important than the competition.

Today, the PS is still the beating heart of the research centre, because it supplies the first accelerating push to the particles which are injected into the LHC ring.

Fig. 2.13 The CERN
accelerator complex

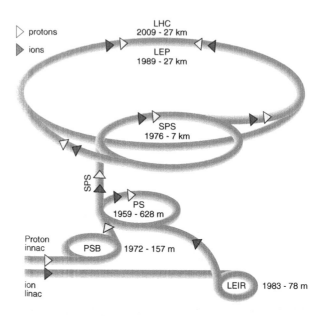

Fig. 2.14 In the CERN
Auditorium John Adams
announces that the night
before the PS accelerated
protons up to 25 GeV
(Courtesy CERN)

The particles accelerated by the PS first enter the SPS and then are directed into the LHC ring with an energy of 450 GeV, as the diagram of Fig. 2.13 shows.

To add a personal story, at this point I should say that I arrived in CERN for the first time a few months later. I also visited the Proton Synchrotron, that I knew from the outset, and the enormous, completely empty, experimental halls where I saw a few experimenters carrying particle detectors in their arms, which were then arranged – it seemed to me with great uncertainty – around the few 'targets' then available. In fact, the PS had functioned much earlier than foreseen, taking the CERN scientists by surprise. Among other things, they did not have the expertise of their Brookhaven colleagues, trained in the USA on many working accelerators of the time. Precisely because of this difference in background, the physics results obtained at the AGS were superior to those from the PS for more than 10 years.

In those few exciting days I decided to apply for a scholarship with the intention of taking an absence from the ISS laboratory for a couple of years. Thus I joined a CERN research group and I worked on a series of experiments mainly dedicated to understanding the properties of antiprotons, the antiparticles of the protons that had been discovered 5 years before with the Berkeley Bevatron.

The 25 GeV protons extracted from the PS were guided by a 'transport line', built of bending magnets and of quadrupoles, and they struck a fixed target, which could be a piece of carbon or a container with a specific fluid inside. In some experiments the creation of new particles caused by the collisions of the protons with the target was studied; among them were antiprotons. In others the collision products were themselves formed into 'secondary beams' of particles, which would normally not occur in nature; these were antiprotons, pions and muons. Their properties could then be studied bombarding other fixed targets with these second-ary beams.

In 1962 I returned to Rome with my family but I continued to go back period-ically to Geneva to take part in another series of experiments, in which nuclear emulsions and cloud chambers were replaced by a new kind of particle detector: the 'bubble chamber' – developed by Luis Alvarez – which had a very important role in the development of sub-nuclear physics.

In a bubble chamber (Fig. 2.15a) the tracks of charged particles are visualised by means of small bubbles that form in a liquid that is 'superheated', i.e. held at a temperature and pressure such that – at the instant the photo is taken – it is just at the boiling point. In these unstable conditions, the energy of the electric charges freed in the liquid by the passage of a charged particle causes local vaporisation of the liquid and thus the formation of micro-bubbles.

Figure 2.15 shows two photos taken by a camera in a hydrogen bubble chamber. The liquid, held at 250° below zero, is at the same time the fixed target and the means of displaying the trajectories travelled by the particles.

In the figure, the charged particle tracks are slightly curved, due to a magnetic field, which is orthogonal to the plane of the paper; from the curvature of each one it is possible to estimate the energy of the particles, and thus calculate their masses applying the law of conservation of energy. The masses of the new particles created in the second collision between a pion and a proton from the hydrogen are computed to be 0.5 GeV (for the kaon K^0) and 1.1 GeV (for the lambda Λ^0).

The Limits of Fixed Target Accelerator Experiments

When a fast pion strikes a proton at rest in a bubble chamber (Fig. 2.15b), not all the total energy it carries is useful for producing the mass of new particles. In fact, a non-negligible part of the incident particle energy is 'wasted' because, after the

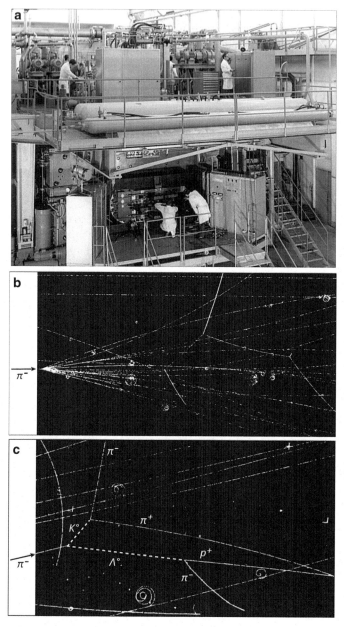

Fig. 2.15 The 2-m hydrogen bubble chamber began to operate in 1967 in a secondary beam from the PS. Below are two events: in the first the collision between an incident pion and a proton creates many particles; in the second two neutral particles are created (a kaon K^0 and a lambda Λ^0 with trajectories indicated by the dotted lines). After a few centimetres both decay into two charged particles ((**a**) Courtesy CERN; (**b**) CERN, Science Photo Library; (**c**) Courtesy Florida State University)

collision, the produced particles continue to move, carrying with them some energy owing to their motion, i.e. their kinetic energy. The production of new particles would be much more copious if the two particles collide one against the other while both are in motion, in the same way that the damage caused by a head-on collision of two lorries is much greater than when a lorry collides with a wall. This is actually what happens in particle colliders, which will be discussed in the next chapter.

To explain in a simple way the energy advantage, consider the collision of a *very energetic* positron with a *stationary* electron, in which the two particles annihilate producing a flash of energy. If the positron travels at 99.99 % of the speed of light, its total energy 35.3 MeV is obtained multiplying the mass ($M = 0.5$ MeV) by the relativistic factor 70.7 quoted on page. Because the stationary electron has only the energy of its mass ($M = 0.5$ MeV), the flash has energy equal to 35.3 $+ 0.5 = 35.8$ MeV; it continues to move, by inertia, in the direction of motion of the positron.

Now suppose that the flash of energy, decaying, produces two particles of equal mass and let us ask what is the maximum value of the mass of each of them. At first sight it might seem that, having 35.8 MeV available, the mass of each of the particles could equal 17.9 MeV. However this is not true because the two particles cannot be produced at rest; the flash is in motion and, by inertia, the moving particles also travel with a non-negligible kinetic energy, which reduces the amount of energy available for the creation of their masses.

To determine the maximum value of the mass of the two particles produced, it is necessary to imagine to 'ride' the energy flash. In this case, Einstein's relativity tells us that the total energy of the flash, *seen from rest*, is given by a formula that is essential in the history of particle accelerators. For incoming particles of large velocities the formula can be written in an approximate and compact form:

$$\text{Available energy in MeV} \; = \; \sqrt{2ME},$$

where the M (mass of the target) and the E (total energy of the projectile) are both measured in MeV.[2]

This energy is available for creation of the two new particles that, from the very special perspective of the stationary flash, depart in opposite directions and with equal kinetic energies. Thus the energy which can be spent on the masses and kinetic energies of the new particles equals $\sqrt{2 \times 0.5 \times 35.3} = \sqrt{35.3} = 5.9$ MeV, much less than the 35.8 MeV which the flash possessed when it was seen in a state of motion. Therefore the two particles that can be produced have masses not greater than half of 5.9 MeV, i.e. 2.95 MeV and *not* half of 35.8 MeV, which would be 17.9 MeV.

---

[2] The same formula applies if the available energy, the mass M and the energy E are all measured in GeV, as it is done most often in this book.

The formula therefore tells us that the collision between a very fast moving particle with a stationary one is an inefficient way to create new particles; in the example – because $5.9/35.8 = 0.16$ – only 16 % of the total energy can be transformed into mass and kinetic energy of the new particles. Moreover, the efficiency diminishes gradually as the energy of the positron increases; if it were to have 3,580 MeV, i.e. an energy 100 times greater, because of the square root which appears in the formula, the available energy would be 59 MeV, only 10 times larger, and the efficiency would be reduced to about 1.6 %.

The same formula also holds in the case in which the particles are different one from the other, as in the case of the collision of a fast electron with a proton at rest. In all cases the efficiency decreases greatly with the growth of the energy that the accelerator imparts to the projectile particle. Indeed, the square root of the formula conveys a grave difficulty in using collisions with a fixed target: to *double* the available energy, it is necessary to *quadruple* the energy of the incident particle.

In the PS, for example, which is 600 m in circumference (Fig. 2.13) and produces protons of 25 GeV, the energy available in the collision with a proton of a target equals $\sqrt{2 \times 1 \times 25} = 7.1$ GeV. If the energy of the proton is doubled, rising from 25 to 50 GeV – with a synchrotron of 1,200 m circumference – the energy available increases from 7.1 GeV to only 10 GeV (because $\sqrt{2 \times 1 \times 50} = 10\,\text{GeV}$).

Of the 50 GeV energy which the proton has acquired in the acceleration cycle, only 10 GeV is available to produce new particles. After the collision with the fixed proton, the final particles retain a large fraction of the original energy, while only a small part is usable for creating new particles.

Is there another less wasteful way to increase the available energy? The answer is yes; the collision between two protons, moving towards one another with equal total energy E, is much more efficient compared to a collision between a moving proton and a fixed target, because all the energy is available to produce new masses and to provide the kinetic energies of the final particles:

$$\text{Available energy} = E + E = 2E.$$

To reach 10 GeV, each of the two protons must have an energy of only 5 GeV. To achieve this, the synchrotron which accelerates one of the beams of particles could have a circumference 10 times smaller than a 50 GeV synchrotron and therefore would be 120 m long instead of 1,200 m.[3] This type of accelerator is what we call today a 'collider'.

---

[3] The circumference of the synchrotron, to achieve a certain total energy, is determined by the magnetic field in the bending magnets. If this is assumed to be fixed, usually at the maximum practical value, the size of the ring must increase proportionately with the total energy.

Chapter 3
The Last Fifty Years

Contents

© Springer International Publishing Switzerland 2015
U. Amaldi, *Particle Accelerators: From Big Bang Physics to Hadron Therapy*,
DOI 10.1007/978-3-319-08870-9_3

Daniele Bergesio, TERA Foundation

In a memorable seminar held on March 7, 1960, in the Frascati Laboratory of the Italian Institute for Nuclear Physics (INFN), the Austrian physicist Bruno Touschek proposed the concept of the electron-positron collider, describing both the machine properties and its physics advantages. Within one year electrons and positrons were circulating in the 'storage ring' AdA.

Touschek was a remarkable man whose life was extraordinary. He was born in Vienna in 1921 of a Jewish mother. Because of the Nazi racial persecution, he took his school graduation examination hiding his true identity and sometime later, in 1943, after studying in Rome and Vienna, but being expelled from the university, fled to Hamburg. There he could not register in the university, but he attended courses while also working, without a formal appointment, in the laboratory of Rolf Wideröe, the man whose 1928 paper – describing the first linear accelerator – suggested to the young Lawrence the idea of the cyclotron.

Reading Wideröe's paper on a novel circular electron accelerator – later called the 'betatron' – and discussing with him, Touschek became so interested in particle accelerators as to aid in developing the theory of the betatron and to help construct it with funds from the Reich Ministry of Aviation, whose undeclared hope was to use it to kill pilots of enemy aircraft.

Despite his need to be discreet, he started to visit the Hamburg Chamber of Commerce regularly to read foreign newspapers. The repeated visits drew attention and he was arrested by the Gestapo on racial grounds, imprisoned and later transferred to a concentration camp. However, on the forced march to Kiel he collapsed and was shot by an SS officer and left for dead, but survived. Later, after the war, he worked briefly with Heisenberg and others in Göttingen, then took up a research fellowship in Glasgow and completed a PhD. In 1952 he moved to Rome, where he remained until the end of his life making outstanding contributions to both the realization of the first electron-positron colliders and to theoretical physics. He was also a very gifted teacher, as I know by having attended his course on quantum field theory.

A Turning Point: Particle Colliders

Rolf Wideröe was not only the inventor of the linear accelerator: in 1943 he devised the idea of the particle collider, which he patented in 1953.

However, the first studies of proton-proton colliders, carried out in the United States in the 1950s, brought to light a grave difficulty. Consider two synchrotrons, which intersect as in Fig. 3.1 and in which bunches of protons circulate in opposite directions. Under these conditions it is like throwing two handfuls of rice at each other: the number of collisions at the crossing point is tiny. This remains true even if the bunches contain millions of particles, because the protons, while crossing, pass too far apart from each other.

To increase substantially the number of collisions taking place each second (a quantity which physicists refer to as the *luminosity* of the collider) it is necessary to reduce the transverse size of the bunches and increase the number of particles in each bunch as much as possible.

To overcome this challenge another invention was needed, but in this case the process was slow and full of detours. The real action started in 1955, when a group of young accelerator experts were working in Wisconsin and Illinois under the

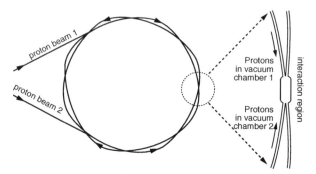

Fig. 3.1 A diagram of the Intersecting Storage Rings, the first proton-proton collider, constructed at CERN in the late 1960s under the leadership of Kjell Johnsen, who had worked in Bergen in the group led by Odd Dahl. Two proton beams with maximum energy of 31 GeV circulated in two synchrotrons each of 1 km circumference, which intersected at eight points

direction of Donald Kerst, who in 1940s, as discussed in chapter 7, had built the first 'betatron'. The young people came from many universities of the Midwest and they are still called the 'MURA group' (Midwestern Universities Research Association).

The problem to be solved was to increase the luminosity by adding new particle bunches to those already circulating in the accelerator without disturbing them. The development of 'particle stacking' required very complicated calculations, which were at the limit of the computers then available. With the limited funds at their disposal, the MURA group decided to build a 50 MeV electron 'storage ring', a synchrotron with particle stacking; this was ready in summer 1961 and, eventually, more than 10 A circulated in the ring of the machine.

Edwin McMillan, who had become head of the Berkeley Radiation Laboratory after the death of Ernest Lawrence, went to MURA's laboratory in Madison to observe the beam during the stacking procedure. He watched for some time, puffed at his pipe and said: "Now *that* is a phenomenon!" (Cole 1994). The MURA experimentalists had achieved what they struggled for, but did not get their construction project approved and the group, unfortunately, dissolved.

In 1971 the first proton-proton collider – built at CERN under the leadership of the Norwegian engineer Kjell Johnsen – had the structure shown in Fig. 3.1; two proton beams of the Intersecting Storage Rings (ISR) circulated for hours in two synchrotrons that crossed in eight 'interaction regions' where detectors were mounted. Since each beam had a maximum energy of 31 GeV, the available energy was up to 62 GeV; in a fixed target experiment such an energy could have been obtained – according to the formula written at the end of last chapter – only if a 2,400 GeV proton were to hit a proton of the target.[1] Due to the very effective way of controlling the stacking of the proton bunches, the final luminosity of the ISR

---

[1] It is normally said that 2,400 GeV is the 'equivalent' energy of a head-on collision between two 31 GeV protons.

Fig. 3.2 Pictures of Bruno Touschek (*left*) and of Gersh Itskovich Budker (*right*), who his colleagues called Andrey Mikhailovich, or simply AM ((**a**) Courtesy of Touschek Family; (**b**) Ria Novosti, Science Photo Library)

was much larger than predicted by Johnsen and collaborators before constructing the collider.

The possibility of a more efficient and elegant collider had been first described 10 years earlier by Bruno Touschek, as we mentioned (Fig. 3.2).

After completing his PhD in Glasgow, Touschek moved to Rome, where he was attracted by the work under way and did not hesitate to remain when offered a position by Edoardo Amaldi. He adapted easily to life in Rome because he had lived there in his youth, as the guest of a wealthy aunt called Ada married to an Italian businessman, with whom she owned a commercial agency.

With his lively Viennese humour, Touschek knew how to capture, with a phrase or drawing, the essence of every situation (Fig. 3.3). His observations sounded particularly amusing in Italian because, while speaking the language very well, he included improbable literal translations from German and English.

The laboratory seminar he held on 7 March 1960 in Frascati has become a milestone in the history of particle accelerators. It was then that Touschek presented the details of an extremely brilliant idea: bunches of electrons and bunches of positrons, each with the same energy, could be injected and would circulate in a *single* synchrotron following *precisely* the same orbit, but in opposite directions, due to their opposite electric charges. In this case the two intersecting vacuum chambers of Fig. 3.1 would not be necessary; one would suffice.

Once circulating in the synchrotron in opposite directions, the electrons and positrons, accelerated contemporaneously by the same radiofrequency cavity – placed at a point along the ring as in a normal synchrotron – could be raised to the desired energy and kept circulating for several hours (provided, of course, that the residual gas inside the vacuum pipe of the ring was sufficiently rarefied not to be an impediment).

Fig. 3.3 When discussions ran on too long, Touschek captured the situation with rapid sketches which, left behind, were collected by friends and colleagues. This one caricatured the debates that were sometimes necessary to determine the field direction of an electromagnet (Courtesy of Touschek Family)

MAGNETIC   DISCUSSION

Furthermore, new electron and positron bunches could be repeatedly stacked along with those already in circulation in the synchrotron; their number would gradually increase (for some minutes, or sometimes for hours) before being accelerated and made to collide at the final energy.

In such a storage ring, at the points of the ring where bunches cross, an electron occasionally annihilates with a positron; their masses and charges disappear and produce a flash of *pure* electromagnetic energy, without electric charge. This flash, of very short duration and twice the energy of each of the original particles, disappears almost instantaneously, being *completely* transformed into rest mass and kinetic energy of new particles.[2]

Bruno Touschek called this new accelerator AdA, which stands for 'Anello di Accumulazione' – i.e. 'Storage Ring', but was also the name of his cherished aunt.

It was clear to him from the outset that the two accumulated and counter-rotating beams should satisfy three conditions: (1) have sufficient intensity; (2) circulate without losing intensity for several hours; (3) have, at the interaction point, transverse dimensions much less than a millimetre, so as to achieve high 'luminosity' and produce sufficient positron-electron annihilations every second. Moreover he well understood that the flash of annihilation energy would be in the form of an

---

[2] This simple and neat reaction was not realized in the electron-electron collider which, in those years, was running in Stanford (California).

energetic photon, which could produce – given enough energy – any pair of positively and negatively charged particles without any complicating effect.

In a short report written in preparation for the Frascati seminar – under the title 'On The Storage Ring' – Touschek mentions the priority of his teacher and friend Rolf Wideröe.

> *The following is a very sketchy proposal for the construction of a storage ring in Frascati. No literature has been consulted in its preparation, since this invariably slows down progress in the first stage. The first suggestion to use crossed beams I have heard during the war from Wideröe, the obvious reason for thinking about them being that one throws away a considerable amount of energy by using 'sitting' targets – most of the energy being wasted to pay for the motion of the centre of mass.*

The story behind the adjective 'obvious' and the patenting of this invention has been written in 1994 by Wideröe, who described Touschek's reaction to him listing, 50 years earlier, the advantages of head-on collisions:

> *He said that they were obvious; the type of thing that most people would learn at school (he even said 'at primary school') and that such an idea could not be published or patented. That was fine, but I still wanted to be assured of the priority of this idea. I telephoned my friend Ernst Sommerfeld in Berlin and we turned it into a very nice and quite usable patent, which we submitted on September 8, 1943.*
>
> *This was given the status of a 'secret patent'. It was not until 1953 that it was retrospectively recognized and published. But we had taken Touschek's objections into consideration and did not state anything about the favourable balance of energy during a frontal collision in the patent, as this was considered a well-known fact. Even so, Touschek was pretty offended.* (Waloschek 1994)

On this occasion Touschek was wrong and his master was right: Wideröe is still recognized as the inventor of the collider concept.

The First Electron-Positron Collider

The construction of AdA began at Frascati immediately following Touschek's proposal and, within one year, was completed, successfully demonstrating the operating principle. AdA was the original ancestor of all particle-antiparticle colliders. It was a small machine, only one metre in diameter, and each circulating beam had an energy of 0.25 GeV (Fig. 3.4).

As soon as it was proven to work it was transported to a French laboratory south of Paris, where electron and positron bunches which contained many more particles than the ones available in Frascati could be produced and injected. The results obtained with AdA prompted a group of French physicists to build a 0.5 GeV storage ring called ACO (Anneau de Collision d'Orsay) in the same laboratory, which started to collect data in 1965. Meanwhile INFN made plans to construct at Frascati a still larger storage ring called Adone – with each annihilation releasing 3 GeV – which began operation in 1968.

Similar rings were built in the same years, and independently of the Italian developments, at Novosibirsk (USSR), where the Russian physicist Gersh Budker

Fig. 3.4 The AdA ring was so small and light that, to alternately inject electrons and positrons coming from the same magnetic beam line, the magnet shown in the picture was rotated every few hours back and forth by 180° around one of its diameters (Courtesy Istituto Nazionale di Fisica Nucleare, Frascati)

proposed both electron-positron and proton-antiproton colliders roughly at the same time as the seminar where Touschek explained his ideas (Fig. 3.2).

Budker was as fascinating a personality as Touschek. The son of a poor Ukrainian peasant killed by a partisan during the Russian revolution, he was so outstanding at school that he gained a place at Moscow University; but the first time he sat the entrance examination, he was rejected for declaring that the devastating food shortage in the Soviet Union was due to the land collectivisation imposed by Stalin's government. He studied for his PhD thesis with the Nobel laureate Igor Tamm and during the Second World War contributed to the Soviet effort developing the atomic bomb, specialising in fission reactors. Then, working with Vladimir Veksler on the design of the Synchrophasotron, he became known as an accelerator expert, proposing a new method for extracting proton beams.

In 1958, already a noted scientist, he was offered the opportunity to direct a new Institute of Nuclear Physics at Akademgorodok in Siberia, then under construction in a pine forest near Novosibirsk. Budker accepted, as a means to withdraw from political and academic quarrels in the capital, which he found hard to avoid because of his stubborn and impetuous character, and within a few years his institute had become the most important in the city of Novosibirsk. In the 1970s, at the peak of its activities, 3,000 people worked there using four accelerators and other facilities designed by Budker for the study of energy production by nuclear fusion. To obtain the money necessary for its research, the institute built, and still builds, electron accelerators used by Russian industry, as well as accelerator components highly prized abroad.

Budker was attracted only by difficult problems and the hardest technical challenges, which he would discuss at length with his senior and junior collaborators, all sitting around an enormous circular table, a sign of democracy unknown in the Soviet Union. Every problem was attacked with care and great optimism, because he used to say: *"The life of an optimist is preferable because he is happy twice; when he plans his work and again when he succeeds. A pessimist can only be happy at most once; if he succeeds"* (Budker 1993).

With his enthusiastic collaborators, Budker designed the VEPP1 electron-positron collider, which was first assembled in Moscow, then transported to Novosibirsk. Its successor VEPP2, whose construction began at the nuclear physics institute in 1960, was completed in 1966 after overcoming enormous technical challenges, exacerbated by Soviet bureaucracy and the shortage of high technology industry.

At the beginning of the 1960s the conditions of work in Italy and the Soviet Union were quite different, but certainly unfavourable. Yet, as we have seen, thanks to two strong personalities, the invention and the first achievement of electron-positron colliders took place outside the United States, which at that time dominated the field both in the size of investment as well as the number of laboratories. Unfortunately, Bruno Touschek and Gersh Budker were not rewarded with the Nobel Prize which, had they been American, would certainly have been the case.

Starting from 1965, storage rings with available energies from 10 to 40 GeV were constructed in Hamburg in Germany, in Cambridge, Ithaca and Stanford in the United States, and in Japan. Finally at CERN in 1989, LEP, a circular collider of 27 km circumference, capable of accelerating electrons and positrons to energies between 50 and 100 GeV came into operation (Fig. 3.5). We will discuss it in detail in Chap. 5.

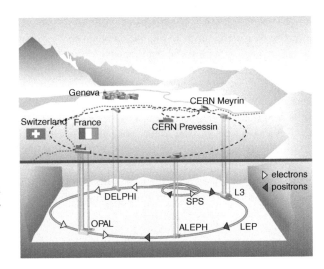

Fig. 3.5 The LEP tunnel, which is 27 km in length and 4 m in diameter, is buried 100 m deep below the plain that extends from Lake Geneva to the foot of the French Jura mountains. The collisions took place in four detectors: L3, ALEPH, OPAL and DELPHI. In 2002 LEP was replaced, using the same tunnel, by the Large Hadron Collider

Proton-Antiproton Colliders

Between the Touschek seminar of 1960 and the storage of electrons and positrons in AdA, less than 2 years elapsed. In contrast, the idea of a proton-antiproton storage ring – proposed by Budker and collaborators in 1966 – had to wait more than 10 years to be realised.

As for electron-positron collisions, a single ring is required to store protons and antiprotons while they circulate on the same trajectory for many hours in opposite directions. In this case however, the difficulty lies in the production of a sufficient number of antiprotons to store. Antiprotons, like positrons, do not exist naturally and must be created in collisions between protons and the nuclei of a fixed target. The mass of the antiproton is 2,000 times larger than the mass of a positron, so its creation requires 2,000 times more energy. Furthermore, it is a rare phenomenon; it is necessary to bombard a target of copper or tungsten with ten million 30 GeV protons to produce a single antiproton!

There is even a further difficulty. Antiprotons emerge from the target over a wide range of angles and energies. This makes it impossible to collect them into small bunches as required to be accelerated in a small cross-section beam pipe like the LEP one (about 5 cm in diameter), in which electrons and positrons circulate for hours.

This obstacle was first overcome thanks to the method of *stochastic cooling*, invented in 1972 by the Dutch engineer Simon van der Meer. *"The cooling of a single particle circulating in a ring is particularly simple"* (Meer 1984), he dared to say in 1984, since *"all"* that is needed is to measure the displacement of the particle circulating in a ring at a suitable location and correct it later with the appropriate electric force, applied when the particle passes through a device (called a "kicker") located at a certain distance (Fig. 3.6a).

But the devil is in the detail. Normally, it is not possible to measure the position of just one particle, because there are so many particles in the ring that a single one is impossible to resolve. So, groups of particles – often referred to as beam "slices" – must be considered instead. For such a beam slice, it is indeed possible to measure the average position with sufficient precision during its passage through a pick-up and to correct its trajectory when the same slice goes through the kicker; one speaks of "stochastic" cooling because the process is statistical, not carried out for each particle individually. In order to do this, very sophisticated electronics is needed, a field in which Simon van der Meer was a master.

For a long time van der Meer spoke about his innovative idea but did not publish, until colleagues obliged him to write a CERN internal note, a document that later did not easily convince the Nobel Committee of his intellectual priority. Luckily the same colleagues had tested the new principle successfully on protons circulating in the ISR and published the positive result quoting the van der Meer note.

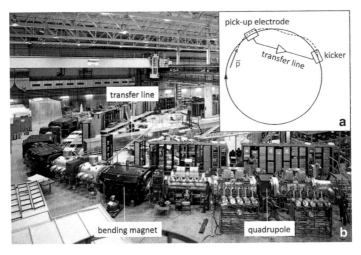

Fig. 3.6 (**a**) To reduce the radial dimensions of a bunch of particles circulating in a synchrotron, a signal is sent from a measuring device to a kicker that corrects the radial position of the particles. (**b**) To contain the wide antiproton bunches, the beam pipe is very broad, as shown by the dimensions of the bending magnets and quadrupoles (Courtesy CERN)

In 1976 the Italian physicist Carlo Rubbia – together with P. McIntyre and D. Cline – proposed to use this technique to construct a storage ring for antiprotons, called the *Antiproton Accumulator* (AA), which began operation at CERN in 1980. It was a very bold proposal, which was initially opposed by the most influential CERN accelerator experts. I remember sitting beside Kjell Johnsen – who had successfully built the ISR, the first proton-proton collider – in the meeting where Carlo Rubbia presented his idea to use the van der Meer method to inject small bunches of antiprotons into the CERN Super Proton Synchrotron (SPS), transforming it into an antiproton-proton collider. At the end Johnsen said to his neighbours: "*I will never believe it will work on the basis of a bunch of computer outputs waved by Carlo*".

The Executive Director-General John Adams was also initially very sceptical. However Rubbia was undeterred and in 1978, with the support of the Research Director-General Léon Van Hove, gained approval for rapid realization of a prototype of the Antiproton Accumulator (called ICE for "*Initial Cooling Experiment*") and, after its successful completion, for construction of the final version shown in Fig. 3.6.

As shown in this figure, to allow the injection and circulation of a sufficiently large number of antiprotons, the beam pipe cross-section of this very special synchrotron is very broad and requires especially large, and therefore heavy, bending magnets and quadrupoles. During the storage phase, the antiprotons circulate as a bunch in the ring for many hours. Once accumulated, they are gradually compressed ('cooled') by adjusting their positions with various kickers. One of the

Fig. 3.7 October 1984:
Carlo Rubbia and Simon
van der Meer celebrating
the announcement of the
Nobel Prize (Courtesy
CERN)

measuring signals is transmitted from one part of the ring to another via the cable labelled 'signal transfer' in the figure. The antiprotons stored in the AA were then, after few hours, injected into the Super Proton Synchrotron (SPS), thus transforming it into a proton-antiproton collider (*Super Proton-Antiproton Synchrotron*).

The first major discoveries made with the new machine were announced in 1983, and earned Rubbia and van der Meer a Nobel Prize: first the W boson, which is responsible for radioactive decay, then the Z. As we will see in the next chapter, these two particles are heavy relatives of the photon, with masses of order 100 GeV (Fig. 3.7).

Stochastic cooling was therefore a technological advance at the same level of importance as the other major steps, which gave rise to the enormous progress in particle accelerators. In summary, these are: (1) multiple acceleration, (2) phase stability, (3) strong focusing, (4) particle stacking, (5) electron-positron colliders and (6) stochastic cooling.[3]

---

[3] Another method (electron cooling) to squeeze the dimensions of antiproton bunches was invented by Budker and tested for the first time in 1974, in the Nap-M storage ring in Novosibirsk. This method is at present widely used.

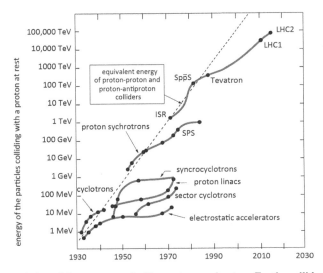

Fig. 3.8 The evolution of the energy reached by proton accelerators. For the colliders (ISR, Spp̄S, Tevatron and LHC) the 'equivalent energy' is plotted, which is the beam energy that would be required in a fixed target collision with a proton to achieve the same available energy obtained in the collider. 'Linac' stands for *linear accelerator*

Past, Present and Future Accelerators

Figure 3.8 (called the "Livingston plot", after Stan Livingston who was the first to draw it) gives an impressionistic summary of the development of proton and antiproton accelerators, from their beginnings in the 1930s to the most recent colliders. It shows how new types of accelerator gradually appeared on the scene over the history of the physics of nuclei and particles. The year is plotted on the horizontal axis and equal intervals on the vertical axis correspond to an increase in energy by a factor 10. Between 1930 and the 1990s the maximum energy of proton-proton (and proton-antiproton) collisions has risen following the dotted straight line, which exhibits *exponential* growth (Panofsky 1997).

In the past the energy reached by specific types of accelerator has not always followed the exponential trend; for example in Fig. 3.8 the evolution of proton synchrotrons, used to bombard a fixed target, diverges from the dotted line. Overall, however, the steady progression has been maintained because new, more advanced and less costly, types of accelerator have supplanted those of previous generations. So it was when the European ISR (Fig. 3.1) and the American Tevatron – the superconducting proton-antiproton collider at Fermilab, near Chicago, which began operation in 1988 – succeeded earlier machines.

The equivalent energy of the *Tevatron* (to be discussed in Chap. 6) is still on the straight line describing exponential growth, while the *Large Hadron Collider* (LHC), which started at CERN in 2009, has a 'delay' of 7–8 years compared to the trend.

The LHC is the proton-proton collider that superseded LEP in the underground 27-km circumference tunnel. It is made of two synchrotrons of equal radii, which intersect at the points where four large detectors are placed to observe the proton collisions. The 'luminosity' of LHC, i.e. the number of proton-proton collisions per second, is definitely greater than in the Tevatron, because the numbers of circulating protons in the two rings are large and equal, while in a proton-antiproton collider the antiproton beam – which circulates in the same ring but in the opposite direction of the proton beam – is very feeble.

As discussed in Chap. 6, in 2009–2012 beams of 3,500–4,000 GeV energy circulated, reaching 7,000–8,000 GeV of available energy. This is indicated by the symbol LHC1 in Fig. 3.8. In 2015 the total energy will be increased to 13,000 GeV (LHC2).

The description and motivation for the LHC will be postponed to the next chapters; here, however, it is important to stress that, despite the enormous energy, the LHC1 and LHC2 do *not* follow the exponential trend represented by the dotted line of Fig. 3.8. To get back to the spectacular progress to which physicists of the twentieth century became accustomed, it will be necessary to devise completely new technologies, based for instance on lasers, which at present are still in their infancy.

Waiting for these long sought but not yet realized developments, in 2014 CERN launched the design study of a Future Circular Collider (FCC) having a circumference of about 100 km and producing proton-proton collisions with available energies in the range 80–200 GeV, which are 6–15 times larger than the energies of LHC. Even if the CERN member States will decide to finance it, the construction of this machine will not start before 2040, since the laboratory is already committed to obtain a factor 10 increase of the LHC luminosity, which should be achieved around 2025.

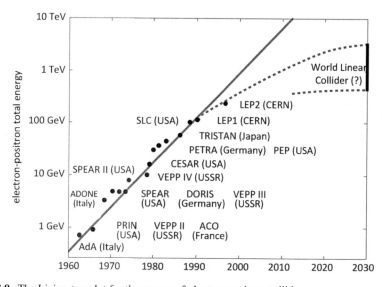

Fig. 3.9 The Livingston plot for the energy of electron-positron colliders

Figure 3.8 describes accelerators of protons and antiprotons, both fixed target and colliders. Figure 3.9 shows instead the analogous behaviour of electron-positron storage rings where, in this case, the ordinate directly represents the available energy (Panofsky 1997). For these storage rings the exponential increase of energy available in the collision was maintained up to the LEP machine at CERN, but has not continued into the twenty-first century. In fact, the hope to continue this growth must be abandoned because of the large energy losses to which such accelerators are subject. A new, seventh, development is absolutely essential, to be added to the six listed above.

Linear Colliders

Electrons and positrons, which have small masses (only 0.5 MeV = 0.0005 GeV), emit X-rays continuously when following a circular orbit. These losses, known as *synchrotron radiation*, increase enormously with electron energy and must be restored on each turn by the radiofrequency cavities, which accelerate the circulating particle beams.

In AdA (Fig. 3.4) and in VEPP2 the electrons and positrons had very low energies and therefore the synchrotron radiation loss was negligible, but at LEP it was necessary to restore *on every turn* 3 GeV of the 100 GeV energy to the circulating particles. The loss of energy by electrons and positrons decreases when the radius of the synchrotron in which they circulate is increased. To limit the maximum loss to 3 % per turn, the radius of LEP was chosen to be 4.3 km and, therefore, a circumference of 27 km. To be able to reach 1,000 GeV losing only 3 % of the energy on each turn, the ring should have been 30,000 km long.[4]

However physicists would like to produce and study electron-positron collisions at energies much greater than the 200 GeV of LEP. This is because – as I will explain in Chap. 5 – it is much easier to understand what is happening reconstructing the trajectories of particles created in simpler electron-positron annihilations than in LHC collisions, where two protons having a diameter of about 1 fm collide, which are, in their turn, made of other particles. To reach this objective *linear colliders* were invented in the 1970s. They are based on a very simple idea: since in storage rings electrons and positrons emit X-rays because their trajectories are curved, to avoid such losses completely it is 'sufficient' to straighten them, by building *two linacs* that shoot electron and positron bunches against each other.

I myself published a paper in 1975 in which I proposed the use of two superconducting linear accelerators which – operating at 270° below zero – would accelerate electrons and positrons up to 150 GeV, colliding them with 300 GeV of available energy. In the same years, Budker and others proposed using linear

---

[4] This problem does not exist for proton colliders, because a proton is 2,000 times heavier than an electron, and the energy lost to X-rays diminishes dramatically with the moving particle mass.

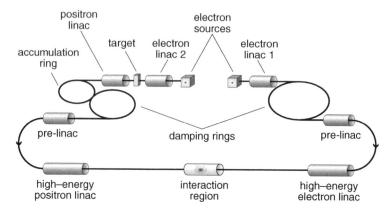

Fig. 3.10 The CLIC project: two linear accelerators – each about twenty kilometres long – "shoot" a beam of electrons at a beam of positrons. The electron bunches are accelerated by "electron Linac 1" and squeezed in the storage rings called "damping rings"

accelerators at room temperature, thus wasting some power but being easier to build. Since then different ideas have been proposed, improved and abandoned.

For more than 20 years at CERN about fifty engineers and physicists have been working on the design, and constructing prototypes, of the *Compact LInear Collider* (CLIC), Fig. 3.10, an electron-positron machine that can reach 3,000 GeV energy in a distance of about 40 km. The two linacs work at a frequency four times larger than the one introduced by Bill Hansen in the 1950s, so the transverse dimensions of the accelerating structure are about four times smaller than those of the accelerator shown in Fig. 2.2b.

In parallel with the development of CLIC, European, American and Japanese scientists have been designing and planning – without defining the construction site – the *International Linear Collider* (ILC), a machine based on superconducting linacs which poses smaller technical challenges and consumes less power but should be three times longer than CLIC to reach the same energy. In the last decades the proponents of the two projects have hotly debated advantages and disadvantages of the different schemes, well knowing that – since the cost will exceed five billion Euros – only one of the two will be, built, if any.

To move towards a common project that would be accepted by *all* high energy physics laboratories and unanimously proposed to the governments of the world, in 2012 the communities working on CLIC and ILC formed the Linear Collider Collaboration (LCC) led by Lyn Evans, the past LHC director (Fig. 3.11b). The aim is to coordinate and advance development work in a global effort to build a single linear collider which would complement the Large Hadron Collider in the future.

In March 2013 Evans said: "*Now that the LHC has delivered its first and exciting discovery, I am eager to help the next project on its way. With the strong support the ILC receives from Japan, the LCC may be getting the tunnelling machines - to make the needed tens of kilometre long tunnel - out soon for a Higgs factory [at 125 GeV + 125 GeV = 250 GeV] in Japan, while at the same time pushing*

Fig. 3.11 Giorgio Brianti, Lyn Evans and Steve Myers (from left to right) have been the leaders of the LHC project during the initial conception (from 1980 to 1992), the final design and construction stages (from 1992 to 2009) and the first running phase (2010–2013) (Courtesy CERN)

frontiers in CLIC technology" (Evans 2013), so as to reach energies larger than 250 GeV within a further decade. For this reason in Fig. 3.9 I have, maybe optimistically, indicated that an electron-positron collider could become operational in 2030, with a total energy of between 500 and 3,000 GeV.

What will be the scientific objectives? To define them in detail it is necessary to await results from LHC2, but it is already clear that the focus will be on detailed study of the Higgs particle and of the hypothetical 'supersymmetric' particles, which I describe in Chaps. 5 and 6.

The Stanford Linear Collider

While in the past all types of accelerator – cyclotrons, synchrotrons and linacs – were built by gradually increasing their size and energy, the plans of the Linear Collider Collaboration foresee, as a first step, the construction of two collinear 250 GeV superconducting linacs which will be about 10 km long in total. It seems an enormous step to undertake in only one stage, but there is a precedent: the 50 + 50 GeV "*SLAC Linear Collider*" (SLC), which can be considered the parent of the future world electron-positron collider, was built at Stanford in the 1980s in a similar forward leap (Fig. 3.12).

SLC was the brainchild of Burton Richter, who in November 1974 was leader of the group that discovered a completely unexpected particle with the electron storage ring SPEAR, that he conceived, directed and brought to completion. During precisely the same period, the discovery of the same particle was also made at Brookhaven, where a group directed by Samuel Ting was following a completely different line of research. This particle – called 'J' by Ting and 'ψ' (read as 'psi') by Richter – was soon understood to be a bound state of a new type of quark (the charm quark) with its antiquark. This discovery was so important for the establishment of

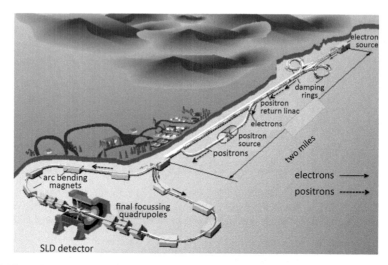

Fig. 3.12 In the SLAC Linear Collider between 1989 and 1998 bunches of electrons and positrons were accelerated in the already existing 2 mile linac of the Stanford Linear Accelerator Centre and then transported by two semi-circular magnetic beam lines to collide head-on at the centre of the 'Stanford Large Detector' SLD (Courtesy of SLAC National Accelerator Laboratory, Stanford University)

Fig. 3.13 For the discovery of the J/ψ particle Burton Richter (*left*) and Samuel Ting (*right*) obtained the 1976 physics Nobel Prize. This picture of 'Burt' and 'Sam' – as they are called by their friends – was taken at SLAC 10 years after the November Revolution (Courtesy Joe Faust, SLAC National Accelerator Laboratory)

the Standard Model of particle physics – which I describe in the next chapters – as to be dubbed 'the November revolution' (Fig. 3.13).

Richter conceived the SLC in 1978, 4 years after the J/ψ discovery with two very precise objectives in mind: to demonstrate a prototype of future linear colliders and to compete with LEP at CERN, in the study of various phenomena observed in electron-positron collisions at 100 GeV of available energy. It was a great success on both fronts.

Here I pause to discuss a unique peculiarity of the SLC: its new way of obtaining the necessary "luminosity", or production of a sufficient number of electron-positron collisions every second, to complete experiments in not too long a time.

At LEP particle bunches circulated for hours, crossing about 100,000 times per second at the points where the detectors were located; so, every few seconds, an electron annihilated with a positron and the many newly created particles were recorded by the four large experiments in Fig. 3.5. In contrast, at the SLC each bunch of positrons and electrons was used only once and the collisions at the centre of the SLAC Large Detector occurred only a 100 times per second.

I already explained that in a collider the particle bunches which meet head to head are similar to two handfuls of rice hurled at each other. In the case of a linear collider what comes to mind is two shotguns – the two linacs – which are fired towards one other. To increase the number of pellets which strike one another each second, it is necessary to reduce the transverse dimensions of the two clusters of pellets – in the region where their trajectories cross – and, at the same time, to fire many more times a second.

Because at the SLC the number of collisions per second was a 1,000 times lower than that at LEP, the challenge was multiplied right from the beginning; to reduce the width of electron and positron bunches by a factor of about a 1,000, and to control their transverse position so accurately that they did not miss the target of the opposite beam. After enormous efforts Richter and his collaborators fully succeeded, achieving collisions between particle bunches with transverse dimensions of about 5 μm (ten times less than a human hair). This technological feat opened the way to a future world collider where particle bunches must have dimensions 100 times smaller to compensate for the fact that, as the energy increases, the annihilation probability decreases.[5]

At the start of the new millenium Richter, who had become SLAC Director, was very concerned about energy and global warming and made important contributions to the understanding of these two very important issues. His book "*Beyond Smoke and Mirrors. Climate Change and Energy in the 21st Century*" – published in 2010 – has become a reference, combining clarity and completeness. Among his many duties, Burt Richter also became President of the *American Physical Society*. Given the motivations that underlie this book, in anticipation of the last two chapters where I discuss the spin-offs of particle accelerators in tumour treatments, it is appropriate to close this section citing something he said in 1995 in his retiring Presidential address:

---

[5] This property is a consequence of the uncertainty principle, which is discussed at the end of this chapter.

The road from science to technology is not the broad, straight highway that many would like to believe. Today's technology is based on yesterday's science; today's science is based on today's technologies. Basic discoveries are at the heart of the development of new technologies, but there are many twists and turns in the road before industrial applications are realized. (Richter 2002)

Richter described the process as a double helix, similar to that of DNA, in which the two strands of science and technology are inextricably linked.

Policy makers in government who think that focusing on short-term applied work can increase economic competitiveness ignore at their peril the implications of this double helix for long-term development. (Richter 2002)

As will become clear when we review the years from 1895 to 2020 in Chaps. 7 and 8, describing the developments of accelerators used in medicine, the history of particle accelerators is a magnificent example of this forward movement along two interconnected spirals.

Electron-Proton Colliders

Electrons, positrons, protons and antiprotons are the only stable particles which, being electrically charged, can be both accelerated by an electric force and deflected by a magnetic field.[6] It is therefore possible to envisage collisions between all these types of particle; however, until 1992 collisions between electrons or positrons with protons had not taken place.

Despite many accelerator projects, the first and only electron-proton collider ever built is HERA (Hamburg), in which between 1992 and 2007 27 GeV electron (and positron) bunches encountered 920 GeV proton bunches, producing electron-proton (and positron-proton) collisions which provided energy available for new particle creation of about 300 GeV. HERA was conceived and strongly driven by an extraordinary Norwegian physicist, Björn Wiik, who studied nuclear physics at the University of Darmstadt and then, from 1965, worked for 7 years at SLAC in California, where he became a good friend of Burton Richter.

In summer 1971 a workshop was held at SLAC dedicated to the design of an electron-proton collider, in which the particles, of 15 and 70 GeV respectively, circulated in two intersecting storage rings. The same year Wiik proposed a slightly larger machine which should be built at the German laboratory DESY in Hamburg. In subsequent years other designs were undertaken in the USA and in Europe, until in 1977 Wiik proposed to add a storage ring for electrons and positrons to the SPS at CERN, which had by then accelerated protons up to 450 GeV for 3 years. Because at that time I was occupied both with neutrino-proton collisions – which are analogous to electron-proton collisions – and with electron-positron linear colliders, Björn involved me in the design of this new collider and from then on I

---

[6] It is not possible to construct accelerators for neutrons or neutrinos because they do not have an electric charge.

a   **b**

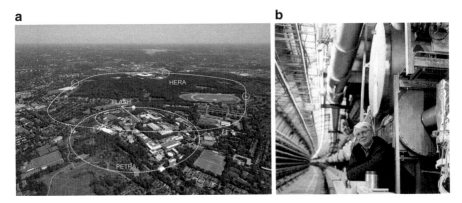

Fig. 3.14 (**a**) In DESY, PETRA was the electron-positron collider where the gluon was discovered by a group of which Wiik was the co-leader (Courtesy DESY, Hamburg). (**b**) Björn Wiik led the DESY group in the construction of the HERA proton storage ring seen in the photo (DESY/ Science Photo Library)

worked with him, and with many other colleagues, towards the goal of building in Europe the first electron-proton collider.

In 1978 the 'European Committee for Future Accelerators' made me chairperson of its 'Working Group on Electron-Proton Collisions'; with the contagious enthusiasm of Wiik, we engaged the support of hundreds of European physicists and engineers. A first design used an electron-positron storage ring called PETRA already in operation at DESY, but was not approved by the DESY Directorate because, even though it was not very expensive, the energy was not sufficiently high. Then the Working Group – in close collaboration with the DESY physicists and engineers – planned a completely new machine, actually HERA, which had to be built at the DESY laboratory too and was approved by the German government in 1984.[7] Wiik was made responsible for construction of the proton storage ring – using those superconducting magnets and a 6.3 km circumference (Fig. 3.14) – and the project was a great success. In particular, HERA experiments brought full understanding of the composite nature of the proton, which I shall describe in Chap. 5.

Wiik was a very quiet man of enormous drive. He always spoke with a soft voice and explained his ideas slowly but very firmly and convincingly. Gunther Wolf, a life-long friend and collaborator, wrote about him.

He had his own style of leading large projects. In filling key positions, he took great pains to find the right people. He was helped by the fact that he had developed special antennae for sensing the qualities and promise of prospective candidates. Once a candidate had been installed in his new position, Björn Wiik would tend not to interfere with his work, even when things went not so well. This trust gave the people working with him room to grow and to gain self-confidence.

---

[7] The project was approved following the offer made by INFN, presided over at that time by Antonino Zichichi, to build in Italy, with Italian funding, half of the superconducting magnets.

Perhaps, the most important virtues of Björn Wiik as a leader were vision, courage, the willingness to assume responsibilities and the ability to instil enthusiasm for the activities of the laboratory through all ranks: secretaries, technicians, engineers, physicists. This enthusiasm resulted in exceptional coherence of the DESY staff and had allowed them to face projects of the size of HERA and TESLA. (Wolf 2001)

TESLA was the name given to the superconducting linear collider, which later became the ILC project. But Bjorn Wiik could not enjoy its development because in 1999 he died at 62 in a tragic accident, while cutting down trees in his garden.

Why Accelerators of Larger and Larger Energies?

The energy of particles in the CERN accelerators has increased in 50 years from 25 GeV at the PS to 4,000 GeV at the LHC and fixed target experiments have been replaced by the much more effective collisions between counter-rotating particles. Why do physicists always want to push particles to the highest energies?

The first reason lies in the equivalence between mass and energy. The famous formula $E = mc^2$ says that, if a mass m disappears, it must be replaced by the appearance of an energy E, given by m multiplied by the square of the speed of light. Conversely, if an energy E disappears, a mass m equal to E/c^2 must unavoidably materialise. This means that in a particle-projectile collision with a particle-target, fixed or moving, the available energy can be transformed into masses of new particles created in the collision; the larger this energy is, the larger is the mass of the particles which can be produced.

Increasing accelerator energies makes possible the creation of particles which are not produced at low energies and which do not normally exist in the matter we find around us because, even if generated in naturally occurring collisions of cosmic rays of the highest energies, they decay immediately into more stable particles, leaving no trace behind.

The second reason to accelerate particles to even higher energies is that it becomes possible to explore smaller dimensions of space and observe greater detail of the subatomic world, as Fig. 3.15 shows.

With the energies reached by HERA and LEP it has been possible to explore dimensions down to a thousandth of the proton radius, which is 1 fm, or 10^{-15} m. With the successor to LEP, the LHC, a ten thousandth of a fermi has been reached. At this scale, as we will see in the next chapter, electrons and other elementary particles – quarks – being much smaller than this – still appear to be point-like objects.

Why do larger energies permit the exploration of smaller dimensions in the subatomic world? The reason follows from the way in which particles interact when they collide.

Consider the example of an electron that, moving close to the speed of light c, collides with a charged particle which, in the case of HERA, was a proton. The collision can be represented by the drawing of Fig. 3.16 – known as a 'Feynman

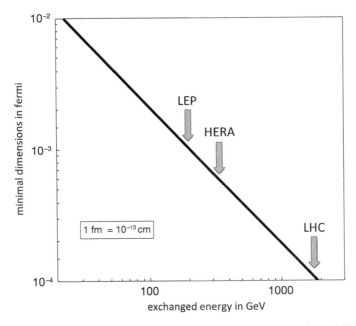

Fig. 3.15 The minimum dimension explored as a function of the energy exchanged. The approximate limits indicated come from experiments carried out at LEP and HERA and, from 2009, at the LHC

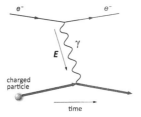

Fig. 3.16 Because of the exchange of a photon of energy E, the electron and charged particle change their directions of motion. In other words, they have experienced a force

diagram' – where the time evolution of the interaction is from left to right. In this representation, before the collision, the two particles approach along converging lines; after the collision they move along diverging lines. This change in direction is essentially caused, in the modern physics picture, by the exchange of a 'force particle' which, emitted by one of the incident charged particles, is absorbed by the other.

In the collision of Fig. 3.16, the two particles interact via the electromagnetic force which is caused by the exchange of a photon (symbol γ). This photon has a special character, because – unlike the photons in a beam of light – it has a very short life and is undetectable by any instrument; for this reason it is called a *virtual photon*.

A virtual photon of energy E lives for the short time allowed to it by Heisenberg's uncertainty principle, which we have previously written in the form $t = 0.2/E$, with E in GeV and t in heisenbergs.[8] For example, when an electron emits a photon with energy equal to its mass, i.e. $E = 0.0005$ GeV, the photon can exist only for a time $t = 0.2/0.0005 = 400$ heisenberg. After that the energy borrowed must be repaid and the photon is reabsorbed by the electron.

During the interval $t = 0.2/E$ the photon can move away from the electron up to a distance given by the product of its speed c multiplied by t. Because a photon travels 1 fm (10^{-15} m) in 1 heisenberg, this distance is $R = 0.2/E$, with R in fermi and E in GeV. In the case of the photon of 0.0005 GeV this means 400 fm which, on the scale of the subatomic world, is a relatively large distance, being 400 times greater than the radius of a proton.

Returning to the example of the collision between an electron and another charged particle, depicted in Fig. 3.16, we note that during the collision the photon energy transferred is distributed for a time t over a volume with radius R.

If the target particle has diameter D larger than R, the photon emitted by the electron interacts with *only a part* of its electric charge, and D can be determined by observing how the electron recoils from the collision. Conversely, if the diameter D is smaller than R, the photon is absorbed by the *total charge* of the target particle and nothing more can be said about D, because the recoil is always the same and independent of the size of the target.

This qualitative reasoning therefore shows that $R = 0.2/E$ (with E in GeV) is the smallest detail – expressed in fermis – which can be studied, in a collision which exchanges a force-particle of energy E. This equation determines the behaviour of the graph of Fig. 3.15, which tells us that – to reduce R – it is necessary to greatly increase the energy available in the collision.

A Third Motive

The two reasons described in the previous section are more than sufficient to justify pushing accelerator energies to higher values. However, another even more compelling motive was discovered about 40 years ago.

In the 1970s, we physicists were intent on using accelerators to observe the production of unstable particles of larger and larger masses, and to study even smaller details of the subatomic world, when – with great surprise and immense enthusiasm – we found ourselves able to retrace the course of time back towards the earliest instants in the life of the primordial universe.

---

[8] At the end of Chap. 1 we called "heisenberg" the time needed by light to cross the diameter of a proton, which is 1 fm. It is a very small duration – only 3×10^{-24} s – but it is chosen because well suited to describe the phenomena of the subatomic world.

When the cosmic plasma was just one microsecond old, collisions took place in which energies of the order of GeV were exchanged. Imagining ourselves going back in time to that moment, we see the universe shrink and the temperature increase. The increase in temperature corresponds to more violent motion of the particles and, therefore, to an increase in the energies exchanged in collisions. For this reason, accelerators of higher energy permit the reproduction of some of the conditions that existed in the remote past ever closer to the Big Bang itself.

So it was that, guided by observation and experimental results, particle physicists and astrophysicists found themselves heading in the same direction: towards an understanding of the history of the universe immediately following the Big Bang. The endeavour, made possible by construction of more powerful accelerators, has grown steadily in extent and in fascination.

Describing the objectives of CERN in popular meetings, I often omit to talk about Heisenberg and Einstein, but I never forget to underline the fact that collisions in the LHC allow us to understand what happened a billionth of a billionth of a second after the Big Bang.

The Epilogue of this book is actually dedicated to this subject, which today is known as *astroparticle physics*.

Chapter 4
Force-Particles and Matter-Particles

Contents

© Springer International Publishing Switzerland 2015
U. Amaldi, *Particle Accelerators: From Big Bang Physics to Hadron Therapy*,
DOI 10.1007/978-3-319-08870-9_4

Daniele Bergesio, TERA Foundation

The CERN control centre resembles those of space missions: a huge room with scores of desks and monitors, small ones on consoles and large ones mounted on walls, displaying real time images, data and plots representing the operation of the various machines. The building is immersed in green foliage on a small road, Route André Lagarrigue. This French physicist is unknown to the general public, but he was at the origin of the first major CERN discovery, so-called 'neutral currents', which opened the way to unification of forces.

In Siena in 1963, during an international conference, André Lagarrigue fell into conversation in a café in Piazza del Campo with Luis Alvarez, the inventor of the radiofrequency linac and the hydrogen bubble chamber. Looking to the future, the two physicists imagined an enormous bubble chamber filled with propane or freon. This would permit the observation of particle-matter collisions, not just of protons and pions but also of the elusive neutrino, which would interact much more frequently with nuclei of those heavy liquids than with the protons of hydrogen.

Subsequently, Lagarrigue succeeded in gaining approval and funding for the new bubble chamber, first in France and later at CERN. It was called 'Gargamelle' after the giant mother of Gargantua, the character created by Francois Rabelais. The name was well deserved, given that it referred to a bubble chamber of 4 metres length and 2 metres diameter. To build it, Lagarrigue, who was immensely gifted as a motivator and organiser, put together the biggest scientific collaboration of the day, made up of fifty physicists and tens of engineers and technicians from six different countries.

The first pictures were taken by the eight Gargamelle cameras at the end of 1970. By March 1972 fifteen frames were selected, which showed very interesting collisions of types never previously observed. Immediately a re-examination of another seven hundred thousand photographs was started, to look for similar events.

The phenomena observed by Gargamelle were the first experimental confirmation of a recently developed theory which aimed to unify the electromagnetic force with the weak nuclear force. This theory, as we will see, has been a great success and today is the foundation of discoveries made at the Large Hadron Collider.

And André Lagarrigue? Unfortunately, he died from a heart attack, following one of the fascinating lectures he used to give at the University of Orsay, Paris, only two years after the discovery. He did not have the satisfaction of receiving the Nobel Prize which he certainly deserved.

From Particle Accelerators to the Discovery of the Subatomic World

In the first three chapters I traced the history of particle accelerators, describing the inventions and technical developments which have taken us from the electrostatic accelerator of Röntgen to the Large Hadron Collider, and paused to reflect on a few of the most interesting personalities who contributed as protagonists in this magnificent intellectual and practical adventure. I chose to talk both of the men – citing their own particularly relevant words – as well as technical subjects; this double choice was suggested by a phrase of Robert Oppenheimer, which struck me strongly when I read it for the first time 40 years ago:

> *We tend to teach each other more and more in terms of mastery of technique, losing the sense of beauty and with it the sense of history and of man. On the other hand, we teach those not destined to be physicists too much in terms of the story, and too little in terms of the substance. We must make more human what we tell the young physicists, and seek ways to make more robust and more detailed what we tell the man of art and letters or affairs, if we are to contribute to the integrity of our common cultural life.* (Oppenheimer 1956)

I move on now to describe the principal results from tens of thousands of experiments – carried out with particle beams from accelerators – and the theories and models which theoretical physicists have constructed to interpret them.

Limitations of space force me to speak little of the experiments in the next three chapters and to present in a schematic way the best theoretical interpretations. The narrative therefore changes pace because I neglect the sophisticated techniques developed by experimental and theoretical physicists and only for a few particularly important arguments – both experimental and theoretical – do I let the leading players speak.

This is in accord with the task I have set myself: to shine a light above all on the contributions of physicists and engineers who invented and built the accelerators and particle detectors, who usually figure only as poor relations in comparison with experimenters and – even more so – theorists. In my opinion they are instead the most important ring in the chain of research in particle physics, as was very well expressed by Victor Weisskopf ("Viki" to everyone), the Austro-American physicist who was Director-General of CERN from 1961 to 1965 and who I remember as the most charismatic of all the DGs in the succeeding years.

> *There are three kind of physicists, namely the machine builders, the experimental physicists, and the theoretical physicists. If we compare those three classes, we find that the machine builders are the most important ones, because if they were not there, we would not get into this small-scale region of space. If we compare this with the discovery of America, the machine builders correspond to captains and ship builders who really developed the techniques at that time. The experimentalists were those fellows on the ships who sailed to the other side of the world and then jumped upon the new islands and wrote down what they saw. The theoretical physicists are those fellows who stayed behind in Madrid and told Columbus that he was going to land in India.* (Weisskopf 1977)

This last statement is particularly surprising because Weisskopf was himself a theorist.

Hydrogen Bubble Chambers

At the beginning of the 1930s, when Lawrence's cyclotron produced the first nuclear reactions, physicists knew of only three 'elementary' particles: the electron, proton and neutron.

In the 1940s other particles, beginning with pions and muons, were discovered using nuclear emulsions to detect collisions of cosmic rays in the atmosphere. Then when the particle zoo began to proliferate in an alarming manner, physicists began to talk of 'fundamental' particles; it seemed impossible that all the newcomers were truly the ultimate elementary components of matter.

The hydrogen bubble chambers, which Luis Alvarez built at the Lawrence Laboratory in Berkeley – named after Ernest Lawrence following his premature death in 1958, were real goldmines of new particles. Bubble chambers had enormous advantages compared to detectors used until then for the study of cosmic rays. Cloud chambers (Fig. 1.9 on p. 19) and nuclear emulsions (Fig. 1.16 on p. 32) did not allow the recording of the many interesting events produced using an accelerator; in a cloud chamber the gas was too thin to have many particles interacting in it and in a nuclear emulsion every event was imprinted for ever without the possibility of erasing it and registering new ones.

Alvarez (Fig. 2.3 on p. 40) had the idea of the hydrogen bubble chamber listening to Donald Glaser, then a young graduate student, who was later awarded the 1960 Nobel Prize. It was a Saturday afternoon; in the final session of the American Institute of Physics conference, the most unconventional ideas were presented to the audience of the few participants who had not yet left.

Glaser described a novel apparatus for particle detection; a container of glass filled with a liquid and equipped with a moving piston, which would allow the fluid to expand to bring it to a pressure and temperature on the point of boiling. Under these conditions the charged particles crossing the 'chamber' would leave tracks made of tiny bubbles, which could be photographed from several directions.

Alvarez grasped the potential of this new instrument and decided to build one filled with liquid hydrogen, which required maintaining it at 250 degrees below zero. The hydrogen would be both detector and target at the same time; the single proton of the nucleus is the simplest target possible. At Berkeley Alvarez brought together the best engineers in the laboratory and within a short time had demonstrated the first bubble chamber.

Over several years Alvarez constructed hydrogen bubble chambers of increasing size – first 2 in. long, then 20 in., finally 72 in. – rather like the progress with Lawrence's cyclotrons. In parallel he organised a complex system to analyse the photographic data; tens of 'scanners' searched for interactions recorded on the film (women researchers, because they were believed to be more careful in repetitive work than their male colleagues) by projecting the images onto large tables, and then measuring the particle trajectories using semiautomatic digitising instruments. The events were finally reconstructed in three dimensions by using the measured projected views.

Within about 10 years, using films from Berkeley and similar bubble chambers in CERN and other laboratories, millions of collisions between fast particles and protons were measured. The result? Hundreds of new particles were discovered, produced in the collisions with the protons in the hydrogen. They were all unstable particles which decayed into lighter particles shortly after their creation.

Protons and Neutrons Are Made of Quarks

Soon physicists found themselves short of letters to represent the particles discovered using bubble chambers, and scientific papers began to be strewn with Greek letters like η, ω, ρ, ϕ, Δ, Σ, Λ, Ξ.

By convention, an exponent denotes the value of the electric charge of the particle, in units of the proton charge, while a bar above the letter denotes the antiparticle; so for example Σ^+ represents the 'sigma' particle, with the same positive charge as a proton, while $\overline{K}^0$ indicates a neutral 'antikaon'.

Fortunately, the proliferation of new 'elementary' particles, almost all unstable, followed a few simple rules; this could be understood from the fact that many

Table 4.1 The three quarks and their antiquarks, which explain the hundreds of reactions observed in bubble chambers in the 1960s and 1970s, have fractional electric charges

Name	Symbol	Charge	Name	Symbol	Charge
d-quark	d or $d^{-1/3}$	$-1/3$	$\bar{d}$-antiquark	$\bar{d}$ or $\bar{d}^{1/3}$	$+1/3$
u-quark	u or $u^{2/3}$	$+2/3$	$\bar{u}$-antiquark	$\bar{u}$ or $\bar{u}^{-2/3}$	$-2/3$
s-quark	s or $s^{-1/3}$	$-1/3$	$\bar{s}$-antiquark	$\bar{s}$ or $\bar{s}^{1/3}$	$+1/3$

reactions, which would create new particles, while theoretically possible from the viewpoint only of energy conservation, in reality did not occur.

It was clear that underlying such confusion there must rein a hidden order, but to discover it particle physicists underwent years of great frustration. Finally, at the beginning of the 1970s, all the pieces of the puzzle fell into place when it was realised that the observed particles were not elementary at all, but rather were composed of other, smaller particles: *quarks.*

Today we know that the many hundreds of particles discovered in bubble chamber experiments during those years are combinations of *just three types* of quark – known by the names u-quark (*up*), d-quark (*down*), and s-quark (*strange*) – and their antiquarks $\bar{u}, \bar{d}$ and $\bar{s}$.

As Table 4.1 shows, the three quarks (and consequently also their antiquarks) have *fractional* electric charges.

The u and d quarks have opposite sign charges; the names 'up' and 'down' reflect these characteristics. Combinations of these quarks make neutrons and protons, and the earliest unstable particles discovered, such as pions. (Antiprotons are made of the corresponding antiquarks listed at the right of the table.) When, in the 1950s, particles now known to contain the s-quark were studied, their average lifetimes appeared *strangely* long; hence the name 'strange'.

However, in 50 years of accelerator experiments not a single particle with an electric charge a fraction of that of the proton has ever been detected; physicists have therefore concluded that, while quarks form composite particles, they cannot exist as single entities outside them. In other words, *free quarks do not exist*; they are bound together in pairs or triplets – as illustrated by Fig. 4.1 – by a new type of force, the so-called 'strong force', and for this reason are called 'hadrons' (from the Greek word *hadrós,* meaning 'strong').

The most common hadrons are the proton p and neutron n which, as Fig. 4.1 shows, are composed respectively of (duu) and (ddu). The total charge of a proton therefore equals $-1/3 + 2/3 + 2/3 = 1$ and that of the neutron $-1/3 - 1/3 + 2/3 = 0$.

Hadrons composed of three quarks are called 'baryons' (from the Greek *barús,* 'heavy'). Protons and neutrons are baryons. Antibaryons are made of three antiquarks; for example the antiproton $\bar{p} = (\bar{d}\bar{u}\bar{u})$ has negative charge $1/3 - 2/3 - 2/3 = -1$.

Combining a quark with an antiquark creates a meson. For example, Fig. 4.1 shows that inside the positive pion, π^+, the strong force – represented by a small

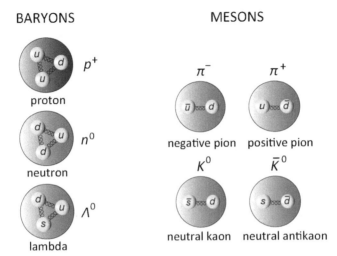

Fig. 4.1 Baryons are made of three quarks bound by the strong force, represented by the springs in the diagrams, while mesons are made of a quark and an antiquark (the images are not to scale: in reality a quark is at least 10,000 times smaller than a hadron)

spring – binds a $\bar{d}$–antiquark (with charge $-\frac{1}{3}$) to a u-quark (with charge $\frac{2}{3}$) forming a particle with charge +1: the $\pi^+ = (\bar{d}u)$

With the discovery of quarks, particle physics changed direction, because the composition of particles in terms of quarks provided simple explanations of the complicated events observed in bubble chambers, in which new hadrons were created.

In some cases there is a simple rearrangement of quarks. For example the collision between a negative pion and a proton can produce a neutral pion and a neutron:

$$\pi^- + p^+ \rightarrow \pi^0 + n^0$$
$$(d\bar{u}) + (duu) \rightarrow (u\bar{u}) + (ddu)$$

The same collision can also produce a more complicated reaction of annihilation and creation of particles. As illustrated by Fig. 2.15c on p. 60, a negative pion and a proton can disappear while their energy is transformed into mass and kinetic energy of a 'kaon' and a 'lambda'[1]:

$$\pi^- + p^+ \rightarrow K^0 + \Lambda^0$$
$$(d\bar{u}) + (duu) \rightarrow (d\bar{s}) + (dus)$$

To understand what happened in this case, it is necessary to examine carefully the quark structure of the individual particles.

---

[1] By definition the hadron containing an $\bar{s}$-antiquark and a d-quark is called a 'kaon' so that the one containing an s-quark is an 'antikaon'.

A $\bar{u}$ antiquark from the pion has annihilated with a u-quark from the proton, and the energy released has created a $s\bar{s}$ pair, following which the $\bar{s}$ antiquark and the s-quark have become bound to the other quarks present, forming a neutral kaon K^0 and the lambda baryon Λ^0.

The American Murray Gell-Mann has been one of the proponents of the 'quark model of the hadrons'. He took the name from a short poem appearing in James Joyce's Finnegans Wake: *"Three quarks for Muster Mark!/Sure he hasn't got much of a bark/And sure any he has it's all beside the mark."* In German 'quark' is a variety of low-fat, soft cheese made from skimmed milk.[2]

The 'quark model of hadrons' has given order to a previously very confused situation, and is today the basis of our understanding; it predicts with great precision the (relatively few) sub-nuclear reactions that can occur, while excluding all the many other conceivable reactions that are not observed in experiments.

Enter the Hadrons with Several Strange Quarks

In this paragraph I depart from the main subject to describe just one of the many experiments – situated like milestones along a road made of failures, discoveries and rediscoveries – which led to full confirmation of the quark model of hadrons.

In 1961 Gell-Mann and, independently, the Israeli Yuval Neeman suggested a new way to classify the then-known hadrons, organising them in regular patterns, the simplest of which contained eight particles (an octet), ten particles (a decuplet) and so on. Because of the basic octets the new and powerful classification scheme was called by Gell-Mann the 'eightfold way', like the path indicated by Buddha in his first sermon: *"Now this, O monks, is noble truth that leads to the cessation of pain. This is the noble* Eightfold Way*: namely, right views, right intention, right speech, right action, right living, right effort, right mindfulness, right concentration."*

The construction was based on a very abstract mathematical symmetry – known as SU(3) – and referred to physical quantities difficult to visualise: 'hypercharge' and 'isospin'. To simplify the explanation, in what follows I will not use the language of the period but the much simpler quark model which Gell-Mann introduced 3 years later.

In Fig. 4.2a, b the first two octets considered by Gell-Mann are shown; they grouped together the then known hadrons which had the lowest masses.

The first contains hadrons made of a quark and an antiquark, which are the three pions ($\pi^- = d\bar{u}$; $\pi^0 = u\bar{u}, d\bar{d}$; $\pi^+ = u\bar{d}$), the two kaons ($\overline{K}^0 = d\bar{s}$; $\overline{K}^+ = u\bar{s}$) and the two antikaons ($K^- = s\bar{u}$; $K^0 = s\bar{d}$). In the figure these are ordered in a way so that (a) the hadrons which are found on the same horizontal level have similar masses;

[2] Independently of Murray Gell-Mann, in 1964 George Zweig, a 27 year old American physicist born in Moscow, proposed a similar model of hadrons made of smaller constituents which he called 'aces' because he thought there were four of them.

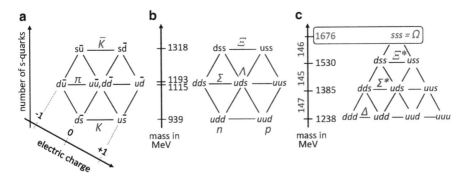

Fig. 4.2 (**a**) The first octet contains, as well as the pion, the kaon K and antikaon $\overline{K}$. (**b**) Together with the neutron and proton, in the second octet are found the lambda (Λ), sigma (Σ) and xi (Ξ), which contains two s-quarks and has a mass of 1,318 MeV. (**c**) The base of the decuplet is the Δ hadron, which does not contain s-quarks and has three charge states: -1 (ddd), 0 (udd), $+1$ (uud) and $+2$ (uuu)

(b) the number of s-quarks increases as one passes, along the vertical axis, from one level to the next above. (In this scheme an anti-s-quark has the value -1, no s-quark equals 0 and an s-quark $+1$.); (c) the electric charge increases when moving from left to right along the inclined axis.

In the octet of Fig. 4.2b the hadrons made of three quarks which have the lowest masses of all the hadrons with the same quark content are grouped together. On the lowest level the nucleons are found (the neutron $n = udd$ and the proton $p = uud$) and on the intermediate level all the hadrons which contain one s-quark are collected, among which we have already encountered the lambda: $\Lambda^0 = uds$. On the same level one finds the three charge states – ($\Sigma^- = dds$), 0 ($\Sigma^0 = uds$) and $+1$ ($\Sigma^+ = uus$) – of the sigma.[3] On the third level of the same octet, the xi particle (Ξ) appears; this contains two s-quarks and has two charge states: Ξ^- (dds) and Ξ^0 (uss).[4]

The eightfold way predicted, as well as octets, multiplets with even more particles, but in July 1962 – when CERN organised its first large international particle physics conference – none of them were known. However, the existence of a hadron which contained an s-quark and had three charge states similar to the Σ of the octet in Fig. 4.2b made of dds, uds, uus, was known. Actually three quarks – for example a u-quark, d-quark and an s-quark – can bind together to form a state with a different mass; a state with a mass that is different from the mass of the Σ^0 belonging to the octet. To distinguish it, this particle is called Σ^*; its three charge states are represented on the second level of Fig. 4.2c.

---

[3] The baryons Λ^0 and Σ^0, which have the same quark structure uds, are different particles because the wavicles of the three quarks have different spatial distributions.

[4] Making use of the charges of the quarks listed in Table 4.1, it is easy to check the charge of the two composite systems: $\Xi^- = dds = -\frac{1}{3} - \frac{1}{3} - \frac{1}{3} = -1$; $\Xi^0 = uss = \frac{2}{3} - \frac{1}{3} - \frac{1}{3} = 0$.

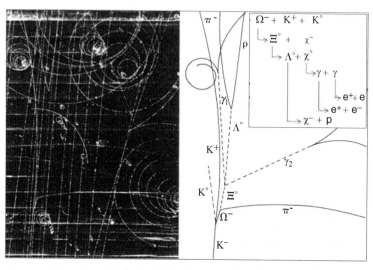

Fig. 4.3 In the photo one sees the tracks of the many negative kaons that traverse the chamber without interacting with the protons of the liquid hydrogen. The tracks belonging to the spectacular event are shown on the right together with the chain of decays that followed the creation of the Ω^- hadron (Courtesy of Brookhaven National Laboratory)

During the CERN conference the discovery of a new particle, the Ξ^*, was announced; it had two charge states and a mass equal to 1,530 MeV. As shown in Fig. 4.2c, the mass difference between the Ξ^*, which according to the eightfold way contained two s-quarks, and the Σ^* was 145 MeV, equal to the mass difference (147 MeV) between the Σ^* and the well-known Δ hadron, which does not contain s-quarks and is shown in the lowest level of Fig. 4.2c.

Murray Gell-Mann understood immediately that this result could easily be explained by a decuplet of his theory – which required the existence of a three s-quark hadron placed at the top of Fig. 4.2c – and suggested the experiment which would allow confirmation of the existence of the new hadron. This singular hadron was named the Ω^-.

Its discovery was described years later by William Fowler and Nicholas Samios, who in the 1970s worked with the large Brookhaven hydrogen bubble chamber and were the leading American players in this story.

> It was now possible to predict with reasonable confidence that the pyramid (of Fig. 4.2c) must be crowned by a single particle at the apex, a particle of negative charge and mass about 1,676 MeV (146 MeV higher than that of the xi particle). In November 1963, Brookhaven undertook a large scale omega-minus experiment. In the ensuing months the entire system – the accelerator, the beam equipment, the bubble chamber, and the cameras – was operated on around-the-clock basis. By 30 January, 50,000 good photos had been obtained. On 31 January there turned up a photograph with a set of tracks that seemed to signal the production of the omega-minus. One could calculate that its mass must have been between 1,668 and 1,686 MeV. This is precisely on the mark for the predicted mass of the omega-minus particle, 1,676. (Fowler and Samios 1964)

The event – reproduced in Fig. 4.3 – was extraordinarily complete and clear. One could easily reconstruct that, in the collision of a K^- ($\bar{u}s$) with a proton, two pairs of *s*-quarks and anti-*s*-quarks were created by the same strong force that binds the quarks to form hadrons. The original *s*-quark of the K^- had joined with the two newly created *s*-quarks to form the omega-minus hadron ($\Omega^- = sss$) while the two anti-*s*-quarks formed the outgoing K^0 ($d\bar{s}$) and K^+ ($u\bar{s}$). After travelling a few centimetres, the omega-minus had decayed into a xi-particle, which had immediately given rise to a lambda (Λ) and a neutral pion (π^0).

The tracks of all these particles were accurately recorded on the film and even the two gammas (γ_1 and γ_2) coming from the decay of the neutral pion were detected through their production of two electron-positron pairs. This double conversion event is very exceptional and it was later calculated that the observed event had a probability of only one part in 1,000 to appear as complete as it did.

The event was discovered on February 9, 1963, by Nick Samios himself during a night scanning shift. The next morning several physicists were crowded around the photo and, as Samios recounted, only then the two electron-positron pairs – crucial for measuring the mass of the omega- were noted.

Someone said, 'Look, there is an electron-positron pair!' And we said, 'My God, that's right!' Someone else said, 'Maybe there is another one from the pi zero!' and someone else said, 'My God there it is!' So there were two electron-positron pairs which I had missed. The significance of these pairs was that they were produced by pi zeros. From the energies of the electron and positron and the angle between them I got the mass of the pi zero and from that the masses of the lambda and the xi! I connected with the other pi and I got a number for the omega which was within 10 % what Murray had predicted. (Crease 1999)

The discovery of this very special Ω^- event convinced everybody of the validity of the eightfold way.

It should be said that, following Gell-Mann's suggestion, a group of European experimentalists also initiated a search for the omega-minus by bombarding the CERN hydrogen chamber with a beam of negative kaons. But it took much more time to set up the beam and to collect a sufficient number of events so the American discovery was announced much before the conclusion of the European experiment. Moreover, the first CERN event was not so beautiful as the very complete and improbable event recorded at Brookhaven. I vividly remember one of the European physicists involved in the experiment repeating the exclamation that was often heard in those days in the CERN canteen: *"Clearly God is American"*.

Matter-Particles

The final column in Fig. 4.4 lists the 'fundamental' particles which make up all matter, both the stable matter we observe around us and the unstable products of high energy collisions, which have very short lives.

To detect these particles and measure their properties has needed four decades and thousands of experiments, carried out at about twenty accelerators and colliders by physicists from all over the world. In the last section we paused to describe one of them. In this and the following three sections I introduce one by one the

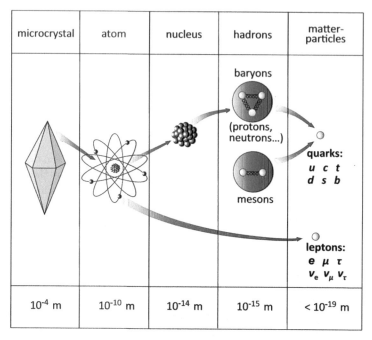

microcrystal	atom	nucleus	hadrons	matter-particles
10^{-4} m	10^{-10} m	10^{-14} m	10^{-15} m	$< 10^{-19}$ m

Fig. 4.4 While atoms in ordinary matter are composed only of *u*-quarks, *d*-quarks and electrons, the many hundreds of known unstable hadrons are made of all six different types of quarks (*d, u, s, c, b* and *t*) and of their antiparticles. In fact there is an antiparticle corresponding to each one of the matter-particles listed on the right. As shown in the fourth column, there are hadrons made of three quarks (baryons) and hadrons made of a quark and an antiquark (mesons)

fundamental particles listed in the last column of Fig. 4.4 and describe the forces acting among them. The facts are many and the presentation is by necessity dry, with no space for the stories of the protagonists of the worldwide intellectual adventure which arrived at the construction of the 'Standard Model' of the fundamental particles and their interactions. The patience of the reader will be hard tested but, I am convinced, the effort will be rewarded when reading on.

In the figure appear not three, but *six* types of quark; as well as the 'light' *u, d* and *s*-of the Gell-Mann quark model, between 1974 and 1995 the *c*-quark (*charm*), *b*-quark (*beauty*) and *t*-quark (*top*) were discovered at accelerators in the USA.

The unusual names of these new quarks have different origins; the names 'charm' and 'beauty' are whimsical, owing to the fascination they provoked in the imaginations of the physicists who predicted their existence. Instead the *top* is the quark which dominates all the others, because it has a mass almost twenty times larger than the already heavy *b*-quark.

The fourth column of Fig. 4.4 shows that the quarks and antiquarks are bound together in mesons and baryons by the 'strong' force, represented by the small springs.

In the final column there are also six 'leptons'; three of these particles are negatively charged (e, μ, τ) while the three neutrinos (v_e, v_μ, v_τ) are uncharged. The leptons, in contrast to quarks, are *not subject to the strong force* but, as we will see, to the 'weak' force (*leptos* in Greek actually means 'weak').

The most well known lepton is the electron which, having negative charge, is often denoted by the symbol e^-. The first neutrino is denoted with the symbol v_e (read as 'nu-e' or 'electron neutrino') because, as we shall see, it has a special relationship with the electron.

Among the charged particles created in collisions of cosmic rays with the atmosphere there is the muon μ ('mu') which has charge -1 (Fig. 1.16 on p. 32); its antiparticle has charge $+1$. Because the electron has a mass of only 0.0005 GeV, the muon (mass $= 0.1$ GeV) is also described as a 'heavy electron' and the other charged lepton, the tau τ ('tau', mass $= 1.8$ GeV) as a 'very heavy electron'.

Like the electron, the muon and the tau also have neutral partners, the two neutrinos v_μ and v_τ (muon-neutrino and tau-neutrino) which complete the list of the leptons.

Coloured Quarks and the Strong Force

In the subatomic world the fundamental forces are the consequence of the exchange of virtual force-particles.

The electromagnetic force between an electron and another charged particle is due to the exchange of a photon, of zero mass, as shown schematically in Fig. 3.16 on p. 85. The electron emits the photon because it possesses an electric charge. The probability of *emission* depends on the value of this charge, which is an intrinsic property of the particle; if the charge is doubled the probability becomes four times larger. The probability of *absorption* of a virtual photon would be zero if the fundamental particle were uncharged, and would quadruple if the electric charge of the particle doubled.

The electric charge is therefore the cause of the electric force.

The same thing happens for all the other forces which act in the subatomic world; every force is due to the emission – and subsequent absorption – of a specific virtual particle (the 'mediator' of the force) and the phenomena of emission and absorption are determined by the value of a specific 'charge' possessed by the particle, which has nothing to do with electric charge.

The small springs drawn in Fig. 4.4 represent the effect of the strong force which binds the quarks inside hadrons. Applying the general principle just mentioned, this force must be due to the exchange of a type of mediator, different from the photon, whose emission probability depends on a new type of charge, which has been given the name 'colour'. These mediators are called *gluons* because they glue the quarks together. Gluons, like photons, are massless and neutral, i.e. without electric charge.

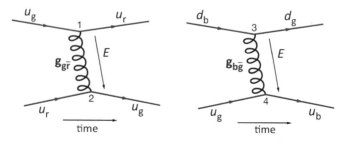

Fig. 4.5 (**a**) At vertex 1 (where three lines join) a green u-quark is transformed to a red *u*-quark by emitting a green-antired gluon, which is absorbed at vertex 2 by the red *u*-quark, cancelling the red of the target quark and changing its colour to green. (**b**) The second diagram illustrates a similar change of colour caused by the exchange of a blue-antigreen gluon. In all cases, the colours of the two final matter-particles are the same as the ones of the incoming matter-particles

In the same way that the emission of a photon is caused by the electric charge of the matter-particle, the emission of a gluon by a quark is due to its 'colour charge', which is carried by quarks but not by leptons (since they do not feel the strong force).

Electric charge is of *two* kinds, which we distinguish by calling them arbitrarily 'positive' and 'negative'. Experiments show, however, that quarks carry *three* different types of 'strong charge', the colour, ensuring that a *u*-quark can be *red, green* and *blue*. So, for example, the three *u*-quarks are distinguished by the symbols u_r, u_g and u_b.

The names should not be taken literally; quarks, point-like matter-particles, are not really coloured! The indices – r, g or b – serve only to distinguish between quarks with the same electric charge and same mass, but which react in three different ways under the influence of the strong force.

The two diagrams in Fig. 4.5 describe elementary phenomena where a gluon is exchanged between two different colour quarks. But here we meet a further complication. There are many types of gluons because the emission of each of them modifies the colour of the emitting and the absorbing quarks.[5] The gluon, which – emitted by a green *u*-quark – transforms it into a red *u*-quark, is different from the gluon intervening when a blue *d*-quark is transformed into a green *d*-quark. As indicated in Fig. 4.5, the diversity of the gluons is made explicit by adding *two* indices to the symbol of the gluon: $g_{g\bar{r}}$ and $g_{b\bar{g}}$. The names of these two gluons are, quite naturally, 'green-antired' and 'blue-antigreen' because in such a way the colours entering at each vertex are *equal* to the colour exiting.

For instance, in the upper vertex of Fig. 4.5a the quark enters with a green colour, while the two exiting particles have red colour (the quark) and green-antired colours (the gluon). This implies that in the vertex the strong force transmits the green

---

[5] In the case of the electric force this does not happen; a negative electron remains a negative electron after emitting a virtual photon, which is electrically neutral.

colour (from the quark to the gluon) and creates two particles (the outgoing quark and the gluon) carrying red-antired colours. Red-antired carries no colour, i.e. it is a 'white' combination and physicists say "the colour charge is conserved".

All these details are difficult to remember, but there is only one really important fact: while all electric forces are due to a *single type* of mediator (the photon, which has no electric charge) the strong forces are caused by the exchanges of *many types* of gluons, each one of them carrying a colour charge defined as red-antired, blue-antigreen, green-antired and so on.

How many gluons are there? Because each one carries a different colour-anticolour combination, one might think that there should be all the possible permutations of three different colours and three anticolours, i.e. $3 \times 3 = 9$. Actually one of the nine colour combinations has to be discarded because it would be white – at variance with the other eight that are coloured –and so cannot transmit the strong force. Thus the gluons responsible for the strong force are reduced to 8; in what follows we will denote them with the symbols g_1 to g_8 so that we do not have to further worry about strange labels like red-antired, green-antired and so on.

All Hadrons Are White

Let us now go back to the hadrons listed in Fig. 4.4. To explain a large number of experimental facts, theorists have been obliged to assume that the three quarks making up baryons (protons, neutrons and lambdas, for instance) always have *three different colours*, which however are continuously exchanged among them. The drawing on the right in Fig. 4.5 shows, for example, that a proton, which initially has the composition (d_b, u_g, u_r) can become (d_g, u_b, u_r), by means of the exchange of a blue-anti-green gluon between the d_b-quark and the u_g-quark.

Therefore in the baryons the strong force, mediated by coloured gluons, binds the three quarks by inducing them to constantly exchange their colours, but always in such a way that, at any moment, red, green and blue are all present.[6]

In electronic screens, red, green and blue are used as 'primary colours' to produce the images; superposing them with different intensities, all the other colours can be obtained. If the three primary colours are superimposed with equal brightness, white is obtained; by analogy it could be said that a proton – despite the continual exchange of coloured gluons between the three quarks – always remains *white*.

---

[6] The strong force, as well as binding the quarks, 'spills over' from the protons and neutrons holding them together in nuclei with a secondary force called the 'nuclear binding force'. It is this residual force which permits the existence of atomic nuclei, made of many protons and neutrons and therefore of the matter which makes up our bodies and every object in the world around us.

According to the theory – called 'quantum *chromo*dynamics' – which describes the strong force and its mediators, it is not just baryons which are white, but also mesons. As shown in Fig. 4.1, the mesons are composed of a quark and an antiquark; white is obtained because the quark carries a colour and the antiquark its anticolour.

Chromodynamics and colour conservation explain very well why quarks are never found outside hadrons, which therefore behave as unyielding prisons.

If, inside a hadron, two quarks (or a quark and antiquark) move away from each other, an exchange of multiple gluons occurs; these, being coloured, attract one another, forming a kind of elastic. The attractive force exerted by this gluonic elastic *increases* gradually as the separation grows; this permanently binds the quarks inside the hadrons by an unbreakable common chain, and prevents them from existing independently.

This is why quarks are never free and how hadrons, which are always white, are made of a quark and antiquark (mesons) or three quarks (baryons). A non-white hadron – made, for example, of four quarks or by five antiquarks - has never been observed.

The three quarks having the same mass and different colours are often said to have the same 'flavour'; so the flavour of the three top-quarks – red, blue and green – is different from the flavour of the three *b*-quarks, while the s_r-quark and the s_g-quark have the same flavour but different colours. (It goes without saying that flavour is only an evocative label that has nothing to do with our taste.)

The quarks inside a hadron continuously exchange gluons, which appear and disappear, giving rise, in turn, to many ephemeral and short-lived quark-antiquark pairs. Quantum chromodynamics describes these complex phenomena but mathematically it is a very complicated theory; it took many decades before theoretical physicists could grasp its sophistications sufficiently well to compute the masses of the hadrons. As shown in Table 4.2, the results of these computations – which require the most powerful computers – are now very satisfactory: the measured masses all fall inside the intervals determined by the calculation.

Because quarks come with six flavours and for each flavour there are three quarks of different colours, the total number of quarks is 18. By adding the 6 leptons of Fig. 4.4, the total number of *fundamental matter-particles* turns out to be *twenty-four*, quite large indeed!

Figure 4.6 summarizes this complicated state of affairs. Since the electric charges and the masses of the three coloured quarks having the same flavour are identical, in the figure the same symbol appears with three subscripts: $u_{r,g,b}$, $c_{r,g,b}$, $t_{r,g,b}$ and so on. Note that each particle has its own antiparticle, which however is not represented here.

Noting the masses, the electric charges and the decays of the 24 types of matter-particle, they can be grouped naturally into three 'families' each of which is made of six quarks (with electric charges equal to $\frac{2}{3}$ and to $-\frac{1}{3}$ of the proton charge) and two leptons (with charge 0 and -1).

Table 4.2 The masses of the baryons are computed by first fixing the values of the masses of the pion (140 MeV), of the kaon (495 MeV) and of the xi-particle belonging to the first octet (1,318 MeV)

Multiplet	Baryon	Quark structure	Measured mass expressed in MeV	Computed mass expressed in MeV
Octet of Fig. 4.2 (p. 97)	Proton p	*uud*	938	935 ± 35
	Lambda Λ	*uds*	1,115	1,110 ± 10
	Sigma Σ⁺	*uus*	1,193	1,175 ± 25
Decuplet of Fig. 4.2 (p. 97)	Delta Δ⁺⁺	*uuu*	1,238	1,190 ± 70
	Sigma star Σ⁺*	*uus*	1,385	1,430 ± 60
	Xi star Ξ⁰*	*uss*	1,530	1,560 ± 40
	Omega Ω⁻	*sss*	1,676	1,675 ± 25

Fig. 4.6 The 24 fundamental matter-particles, 18 quarks and 6 leptons, with (beneath the symbols) the values of their masses in GeV; the electric charge is indicated on the *left*. The figure makes the symmetry between quarks and leptons clear

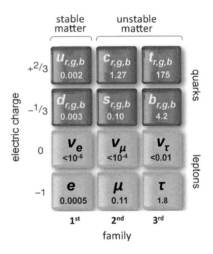

Figure 4.6 also records, under each symbol, the mass of the particle, expressed in GeV.[7] The particles of the first family have masses less than 0.005 GeV, while those of the second family have masses below 1.3 GeV. The heaviest matter-particle, the *t*-quark, is a hundred thousand times more massive than the *u*-quark.

---

[7] The three neutrinos have such small masses that in Fig. 4.6 only experimental upper limits are reported (the symbol < means 'less than'). These insubstantial matter-particles do not have sufficient energy to decay into other particles, therefore they are stable; once produced, for example immediately after the Big Bang, neutrinos live for ever, moving almost at the speed of light and interacting very little with matter.

The charged particles of the second and third families, thanks to their masses, have energy in abundance and decay very quickly into those of the first family; this is why *all* stable matter is made only of charged matter-particles from the first family, *d*- and *u*-quarks (which bind together to form neutrons and protons) and electrons e^-.

How big are these particles? Experiments carried out at the most powerful accelerators have allowed exploration of distances of the order of a ten-thousandth of a fermi (Fig. 3.15 on p. 80). However, even on this scale, all the leptons and quarks still appear to be point-like; so in Fig. 4.2 I have shown their diameters to be less than 10^{-19} m.

Before closing the section it is worth remarking that the masses of the baryons are much larger than the sum of the masses of their constituent quarks. For example, since the masses of a *u*-quark and a *d*-quark are 2 and 5 MeV respectively (Fig. 4.6), the masses of the three quarks contribute *only* 9 MeV to the mass of a proton, which is a hundred times bigger. The reason is that the proton mass is essentially due to the energy of the quarks, gluons and antiquarks that continuously move inside it; quantum chromodynamics describes this incessant activity very accurately, as proven by the comparison of Table 4.2.

The Weak Force

Figure 4.7 illustrates the forces which act on matter-particles.

The *strong force* – due to the exchange of coloured gluons between quarks – is the most potent. It produces two effects: it binds the quarks and antiquarks to form hadrons, as shown in the figure by the two grey arrows, and causes a change of direction by two colliding hadrons, giving rise to a process of 'scattering'.

The *electromagnetic force*, which results from the exchange of photons and acts on charged particles, also produces two effects: the binding of electrons to atomic nuclei and scattering of two charged particles, as in Fig. 3.16 on p. 85. However it is much less powerful than the strong force.

The *gravitational force* is even feebler; when it acts between two matter-particles it is so weak that it has no measurable consequences. Nevertheless the effects of the gravitational force do add up, given that matter-particles that have negative 'gravitational charge' do not exist (unlike the case of the electric force). In fact it is the sum of the effects of the extremely feeble gravitational force acting on the enormous number of particles in celestial bodies which keeps the planets bound to the Sun and confines the stars in galaxies and galaxies in clusters. The mediator of the gravitational force is the graviton, which is still out of reach of experimental detection.

The *weak force*, as its name says, is feebler than the electromagnetic force (but much, much stronger than the gravitational force). As the grey arrows in Fig. 4.7 indicate, the weak force does not create bonds but makes its effects felt in the *transformation* of matter-particles, i.e. particle decays or 'disintegrations'.

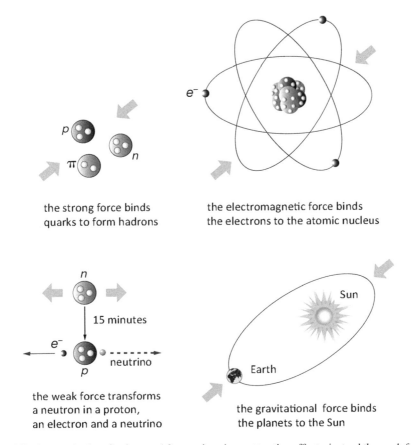

the strong force binds
quarks to form hadrons

the electromagnetic force binds
the electrons to the atomic nucleus

the weak force transforms
a neutron in a proton,
an electron and a neutrino

the gravitational force binds
the planets to the Sun

Fig. 4.7 Among the four fundamental forces, three have attractive effects; instead the weak force does not bind matter but causes the 'decays' of matter-particles

This force is the least known, but is crucial for our life; thanks to it four hydrogen nuclei can merge into a helium nucleus inside the Sun, in a chain of nuclear fusion reactions which release energy and make our star shine (as well as producing innumerable neutrinos).[8]

It is really the feebleness of the weak force which ensures that the Sun 'burns' its nuclear fuel *slowly*; in 4.6 billion years it has consumed less than half of it. Despite that, the production of neutrinos at the centre of the Sun is so abundant that a

---

[8] In nuclear fusion of hydrogen, 2 of the initial 4 protons undergo the inverse process to the one shown in Fig. 4.7 from the effect of the weak force; each proton is transformed into a neutron, a positron and a neutrino. In this way a helium nucleus, composed of 2 protons and 2 neutrons, is produced.

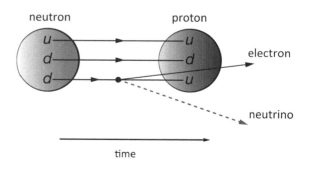

Fig. 4.8 In beta decay of a neutron a *d*-quark is transformed into a *u*-quark emitting a new virtual W⁻ force-particle

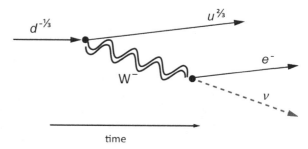

hundred million solar neutrinos pass through the body of each one of us every second.

Figure 4.7 shows the simplest of the effects caused by the weak force; the 'beta decay' of an isolated neutron (which has an average lifetime of around 15 min) into a proton, an electron and a neutrino.

Figure 4.8 illustrates the phenomenon in greater detail; in order for a neutron to be changed into a proton, at the moment of disintegration one of its two *d*-quarks must be transformed into a *u*-quark. As the lower diagram shows, this occurs with the emission of a virtual force-particle called W⁻ ('W-minus'; it must be negative in order to conserve electric charge).

Because of the weak force, neutrons and protons bound in atomic nuclei can also decay into one another, producing electron-neutrino pairs (as in Fig. 4.8) or their antiparticles; this is 'beta radioactivity' which is such a concern in nuclear accidents, but which is also the basis of techniques used for diagnosis of tumours, such as scintigraphy or Positron Emission Tomography (PET), of which we will speak in Chap. 7.

The left diagram of Fig. 4.9 illustrates the phenomenon – exploited in PET – of beta positive decay of a nucleus, which occurs when a proton is transformed into a neutron with the emission of a positron; the mediator is the same as that in the neutron-proton transformation (Fig. 4.8) but carries positive electric charge and is called "W-plus".

Fig. 4.9 The virtual force-particles W $^+$ and W $^-$ cause (on the *left*) the *positive beta decay* of a nucleus, with the emission of a positron and a neutrino, and (on the *right*) *electron capture* with the emission of a neutrino

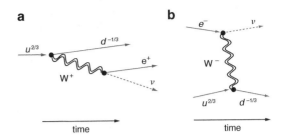

Another typical phenomenon for which the weak force is responsible (through the intermediary W$^-$) is 'electron capture', represented on the right of Fig. 4.9; an atomic electron, which circulates around a nucleus, is absorbed by a quark – found inside the nucleus – and is transformed into a neutrino, while the quark changes its flavour.

The intermediaries W$^-$ and W$^+$ are usually called the 'charged intermediate vector bosons', a long and difficult name. Here I will call them instead the 'charged *asthenons*' (from the Greek *asthenés*, 'without force, weak'); it is a little used name but has the advantage of reminding by assonance that it refers to a fundamental force-particle similar to the *photon* and the *gluon*, in this case mediator of the weak force.

Neutrinos, Neutrettos and the Cold War

In the last pages I have put many irons in the fire: 24 matter-particles – three of which are invisible 'neutrinos' – 4 fundamental forces with their force-particles, quarks which attract one another because they exchange 'coloured' gluons, radioactive decays mediated by asthenons, and so on. Certainly the idea most difficult to swallow is that of the three neutrinos, the mysterious particles whose origin and necessity were not explained. Therefore I will pause and recount one of their many stories.

From the start of the last century physicists knew that electrons emitted in beta decay of radioactive nuclei had a strange behaviour. The laws of physics dictate that, if a nucleus emits only *one* electron, its energy should always be the same; instead experiments demonstrated that identical nuclei emitted electrons of *all* energies, from zero to a maximum value characteristic of the specific nucleus. In the 1930s, to explain this fact, the great Austrian theoretical physicist Wolfgang Pauli invented what he himself judged "an incredible solution" which he felt able to propose because "only who dares can win"; together with the electron a *neutral* particle is created, which has a mass much smaller than that of the electron and carries with it a part of the decay energy.

At that time the leading academic professorship in Italian theoretical physics was occupied by the 33 year old Enrico Fermi in Rome. Fermi understood the potential implied by Pauli's hypothesis and, applying the newly formulated quantum mechanics, explained the distribution of electron energies from radioactive nuclei by supposing that a new force induced the transformation of a neutron into a proton with the emission of a pair of other particles – an electron and a neutral electron called the neutrino, which had not previously existed inside the nucleus. Similarly, according to Fermi's theory, a positron-neutrino pair is created in the transformation of a proton into a neutron inside a radioactive nucleus.

The name 'neutrino' was used for the first time in Via Panisperna in Rome, where Fermi's department was located.[9] The physics community rapidly accepted Fermi's theory and the existence of a new force, which caused the decay of radioactive nuclei with the emission of neutrinos, and was then called 'weak'.

Ten years after Fermi's theory was proposed, the muon (or heavy electron) was discovered and ten years after that, Powell and collaborators observed in their emulsions the decay of a pion into a muon followed by the decay of a muon into an electron (Fig. 1.16 on p. 32). It was therefore quite natural to assume that the neutral particles, which carried away part of the energy and were not photons, were also neutrinos. For this reason in Fig. 1.16 the decays of the pion and the muon are written in the form:

$$\pi \to \mu + neutrino1 \qquad \mu \to e + neutrino2 + neutrino3.$$

In the case of the second reaction, it is necessary to assume that *two* neutrinos were created because the energy of the electron is different in each decay and this can occur only if in the final states there are three particles, as happens in the decay of the neutron $n \to p + e + v$.

Immediately an important debate began: is it possible to observe the neutrinos emitted in beta decay? Is the neutrino1, emitted in the pion decay, the same type of neutrino emitted in the decay of the neutron? Is neutrino3 the same type as neutrino1 and neutrino2?

One who distinguished himself as an active participant in this debate was Bruno Pontecorvo, who in the 1930s was the youngest among Fermi's pupils, born into a prosperous intellectual Pisa family; among his brothers, Gillo became a well known film director, noted especially for "The Battle of Algiers", and Guido – who

---

[9] Pauli called the hypothetical neutral particle the 'neutron', but with the discovery in 1930 of the neutron, which is an essential component of all atomic nuclei, this nomenclature created much confusion. For this reason in 1933 Enrico Fermi began to refer to the new particle as 'neutrino', following an exchange in via Panisperna when Edoardo Amaldi – Fermi having explained that Pauli's neutron was not the one from nuclei because it had a much smaller mass – said without too much thought: "But then that is not a neutron, it's a neutrino". (In Italian the diminutive is expressed by applying the suffix –ino to a noun.)

Fig. 4.10 This amusing comparison between the intuitions of Newton and Pontecorvo was drawn by Misha Bilenky, son of the well-known Russian physicist Samoil Bilenky who was a friend and collaborator of Pontecorvo at Dubna (Courtesy of Mikhail Bilenky)

emigrated to the United Kingdom – became a renowned geneticist and a Foreign Member of the Royal Society.

In 1936, the 23 year old Pontecorvo went to work with the Nobel laureates Irène and Frédéric Joliot-Curie in Paris, where he came into contact with antifascist refugees, which strengthened his communist convictions. From Paris, during the German invasion of France, he daringly escaped by bicycle to reach the south and from there travelled with his family to Oklahoma, where he developed a method for oil prospecting based on the detection of neutrons. In 1943 he was recruited to work in Canada where he contributed to the design and construction of the first reactor in which neutrons were slowed by collisions in heavy water. In this period he realised that, given the large flux of neutrinos produced in such a reactor, even if they interact minimally with matter, it would be possible to detect at least some each year by applying radiochemical methods, in particular using chlorine. Then, in 1946, he underlined that the same method could be used to measure the flux of neutrinos originating in the nuclear reactions taking place in the Sun (Fig. 4.10).

Pontecorvo also related the decay of the negative muon ($\mu^- \rightarrow neutrino2 + e^- + neutrino3$) to the decay of the neutron ($n \rightarrow p + e^- + \nu$) arguing that they were caused by the *same* force – the weak force – and that, as the neutron is transformed into a proton by emitting an electron-neutrino pair, so the muon is transformed into a neutrino2 – which is similar to the muon – by emitting an electron and a neutrino similar to the electron. Convinced that the two neutrinos were different particles, he used to call them 'neutrino' and 'neutretto', which later became known as ν_e (electron-neutrino) and ν_μ (muon-neutrino).

Fig. 4.11 In 1949 Bruno Pontecorvo (on the *left*) and Enrico Fermi (second from the *right*) visiting the Olivetti factories at Ivrea (Courtesy Enriques Family)

Pontecorvo was invited in 1949 to work at the Harwell nuclear research centre in Great Britain, and in the summer of 1950, while on vacation in Italy with his family, he secretly left Rome for the Soviet Union fearful of the possibility that the Korean war would cause a third world war and very conscious of being under observation by British counterespionage as a consequence of his political beliefs (Fig. 4.11). His flight to the East caused a great sensation because it was immediately thought to be connected with the discovery, a few months previously, of spying by Klaus Fuchs, a naturalised British physicist of German birth, who had worked on the American atomic bomb project. In fact Pontecorvo had never worked on any of the Allied military projects, although he well understood the physics of nuclear reactors. Years later he declared that he had never been aware of and had never transmitted secrets connected to the atomic bomb to the Soviets and that he had never worked at nuclear weapon projects in URSS, as confirmed by the content of his laboratory logbooks in Dubna, made public in 2013.

What is certain is that Bruno Pontecorvo continued to cultivate his passion for neutrinos and neutrettos, working both theoretically and experimentally in the Dubna laboratory, where he used the Synchrophasotron built by Veksler and created a prestigious school of physics. Among other things, he suggested the possibility that neutrinos, in contrast to photons, could have a mass, even if very small, and that therefore it was possible that a neutrino in flight could transform into a neutretto. Today we know that this incredible phenomenon, which is known as

Fig. 4.12 The processes examined by Pontecorvo are described by these two diagrams, which differ from those used in 1959 because they are based on today's knowledge of the nature of the weak force and the quark structure of nuclei

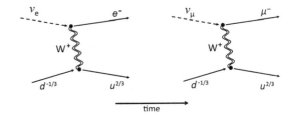

'neutrino oscillation', happens to two thirds of the neutrinos which, emitted by the Sun, pass through the Earth.[10]

Much before this, in 1959, the international particle physics conference was held in the Soviet Union for the first time, in Kiev. A group of western physicists participated and Pontecorvo presented his latest ideas on the neutrino and the neutretto. Essentially Pontecorvo described – in the language of the day – the two processes illustrated in Fig. 4.12 underlining the fact that, if the neutrino v_e is different from the neutretto v_μ, in the collision with a nucleus the neutrino *always and only* gives rise to an electron in the final state, while a neutretto (i.e. a muon-neutrino) produces *always and only* a muon, which is nothing more than a heavy electron.

He suggested an experiment designed to check this hypothesis: an intense beam of protons of many hundreds of MeV would create, striking a metal slab, a flux of pions which after, leaving the slab, would decay – travelling tens of metres in air at almost the speed of light – producing a beam of neutrettos (and *not* neutrinos) and muons. The muons could be stopped by an iron shield a few metres thick; instead the neutrettos, crossing it practically undisturbed, would interact, in small numbers, in a detector in which they would produce muons, and *not* electrons.

Less than 6 months later the same experiment was independently proposed by a 27 year old American physicist working at Brookhaven National Laboratory. Melvin Schwartz wrote about how it all came about.

One Tuesday afternoon in November 1959, I happened to arrive at Columbia late at coffee to find a lively group around T.D. (which is what we called Tsung Dao Lee, a famous theoretician) with a conversation ensuing as to the best way of measuring weak interactions at high energies. A large number of possible reactions were on the board making use of all the hitherto known beam particles – electrons, protons, neutrons. None of them seemed at all reasonable. That night it came to me. It was incredibly simple. All one had to do was use neutrinos. The neutrinos would come from pion decay and a large shield could be constructed to remove all background consisting of the strongly and electromagnetic interacting particles and allow only neutrinos through. Since neutrinos interact only weakly, any interaction arising from them would be, ipso facto, a weak interaction. Some months later I found out that Professor Bruno Pontecorvo had proposed that neutrinos could be used as probes of the weak interactions during a conference in Kiev in 1959.

---

[10] The first observation was due to the American physicist Raymond Davis, who applied the Pontecorvo chlorine method (figure 58b) and was awarded the 2002 Nobel Prize for the experimental detection of solar neutrinos.

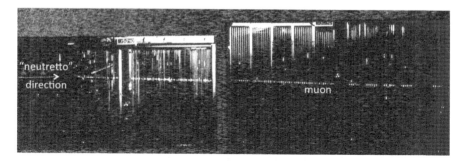

Fig. 4.13 In the BNL experiment the neutrinos came from the *left*. The long track, formed by many bright sparks, visualizes the trajectory of a muon that has great penetrating power. The shorter track is a hadron, probably a proton, set in motion in the collision of the neutretto with an atomic nucleus of one of the spark chambers (Courtesy of Brookhaven National Laboratory)

> *Apparently the significance of his remark had been missed by the few Americans who were present and their substance was never transmitted back.* (Schwartz 1972)

At Brookhaven National Laboratory in the USA, Leon Lederman, Melvin Schwarz and Jack Steinberger, who were rewarded by a Nobel Prize in 1988, completed in 1962 the experiment which, to his great disappointment, Bruno Pontecorvo could not carry out at Dubna because of insufficient instrumentation.

For many months the high-intensity 30 GeV proton beam from the Alternating Gradient Synchrotron (AGS) produced a large number of pions, which travelled 20 m at almost the speed of light towards a 5,000-t steel wall made of old battleship plates. On the way, they decayed into muons and neutrinos. The steel wall stopped the muons and the residual pions, while about 10^{14} muon-neutrinos traversed the detector, a 10 t system of neon-filled 'spark chambers' in which the passages of charged particles were visualized as sequences of aligned sparks. The cameras recorded 29 events similar to the one reproduced in Fig. 4.13; the very long tracks observed were due to penetrating muons *and not* to electrons, which would have left a cluster of much shorter tracks. This proved that the neutrinos produced together with a muon are indeed 'neutrettos' – different from the electron-neutrinos created in nuclear decays; when hitting a nucleus of the spark chamber, they always transform into a muon, as in Fig. 4.12b, and never into an electron.

Since then we know that neutrinos are different from neutrettos and that the weak force involves the pairs (e, v_e) and (μ, v_μ) to which the W asthenon couples equally. When the tau, another heavy electron, was discovered, it was natural to expect that it also should have a neutral counterpart, later discovered, the tau-neutrino. Thus, this led to three neutrinos.[11]

---

[11] Since in Italian it is easy to invent diminutives, augmentatives and pejoratives by using modifying suffixes, I dare to suggest for the three neutrinos a new, compact and meaningful nomenclature: *neutrino, neutretto* and *neutrotto* ('fatter neutrino') are certainly better than

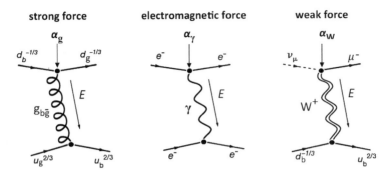

Fig. 4.14 The probability of collisions caused (**a**) by the strong force, (**b**) by the electromagnetic force and (**c**) by the weak force is determined by three couplings α_g, α_γ and α_w

Why Are Some Forces 'Strong' and Others 'Weak'?

To compare the three forces – strong, electric and weak – which act in the subatomic world, let us examine the three phenomena illustrated in Fig. 4.14.

As we know, they are caused by the 'charges' carried by the intervening matter-particle: the 'electric charge' for the electric force and 'colour' for the strong force. To make the argument more quantitative, in the figure I have introduced a new quantity α, which is called a *'coupling'* and represents the probability of emission (or absorption) of a force-particle by a matter-particle.

The electromagnetic coupling – α_γ – of the photon to the electron defines how many times it is necessary to observe the charged particle, on average, before finding it with a virtual photon. This has the value $\alpha_\gamma = 0.0073 = 1/137$, which means that an electron must pass near to another electron 137 times before it emits a virtual photon so that a collision can take place. The phenomenon occurs only about 7 times in a thousand, so it is rare and for this reason the electromagnetic force is faint.

Instead the strong force is intense, because the coupling α_g of the gluons exchanged inside a hadron is of order 1. In other words each quark emits and absorbs gluons constantly, so that the probability of finding one at a distance from a quark of the order of the size of a hadron (i.e. 1 fm) is 100 %.

The diagram of Fig. 4.14c represents the interaction of a muon-neutrino with a quark bound in an atomic nucleus. Because there is a strict relationship between the neutrinos and their leptons, a muon-neutrino (v_μ) which exchanges an asthenon with a quark is always transformed into a muon. The phenomenon is caused by the exchange of an asthenon W, and experiments on electron capture, radioactive decay

---

electron-neutrino, muon-neutrino and tau-neutrino in conveying the fact that they refer to neutral particles of very small and increasing masses.

and the interaction of neutrinos have demonstrated that the coupling has a value $\alpha_W = 0.0016$, and is therefore similar to the coupling of photons to electrons. Note that the absorption of an asthenon, as shown in the lower part of Fig. 4.14c, changes the flavour of the quark but not its colour.

In that case, why is the weak force much feebler than the electromagnetic force?

The fact is that, in contrast to photons, which are massless, the asthenon has a very large mass; as we will shortly see, it has a value of about 80 GeV, equivalent to about 80 protons combined. It is actually this large mass that makes the exchange of a W manifest itself as a force of slight intensity. The reason, as usual, is linked to the uncertainty principle. The muon-neutrino of Fig. 4.14c can emit a particle with a mass of 80 GeV but only for a very brief interval since, as discussed in Chap. 1, the uncertainty principle allows a loan of the energy $E = 80$ GeV from the 'Heisenberg bank', but the greater the amount of energy borrowed the earlier the loan matures.

This short time is easily computed with the formula $t = 0.2/E$, where E is in GeV and the time is expressed in "heisenberg" – equal to 3×10^{-24} s (in Chap. 1 I called this time "the heartbeat of a proton"). In the case of the exchange of a virtual W asthenon, the maximum time is $0.2/80 = 0.0025$ heisenberg, which corresponds to 8×10^{-27} s.

The virtual asthenon, reabsorbed almost immediately by the matter-particle that emitted it, has therefore almost no time to encounter the d-quark and interact. So the weak force is faint not because the coupling is small, but because its action is effective for very short time intervals, which means over very short distances, as a result of the 80 GeV mass of its force-particle.

Charged Currents and Neutral Currents

The collision of a muon-neutrino with a d-quark and its consequent transformation into a muon (Fig. 4.14c) is described by physicists as a *charged current*, because the virtual W^+ force-particle – which transfers energy from one vertex to the other – takes with it an electric charge. This is a characteristic peculiar to the weak force; the mediators of the other three forces – strong, electromagnetic and gravitational – are all neutral, and the phenomena which they induce are therefore always due to 'neutral currents'.

At this point a question spontaneously arises: is there a *neutral* mediator of the weak force with characteristics similar to those of the charged asthenons W^- and W^+?

If such a mediator exists, it should give rise to reactions such as the one illustrated in Fig. 4.15, where a muon-neutrino experiences an interaction with an electron without changing its properties, thus remaining a muon-neutrino.

The search for the phenomenon of *weak neutral currents* lasted for 20 years. It was finally observed by a team of scientists which André Lagarrigue organised to build and operate the Gargamelle bubble chamber, specifically designed for the study of interactions caused by an intense beam of muon-neutrinos, produced at

Fig. 4.15 In analogy with the exchange of a charged asthenon (Fig. 4.14c), physicists wondered if a neutrino can exchange a neutral force-particle, and remain a neutrino. The mediator was called Z^0 from the beginning

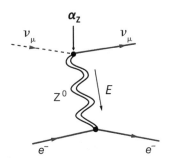

CERN using the method invented by Pontecorvo and Schwartz.[12] Actually, at the beginning the search for neutral currents was not considered very important, mainly because earlier experiments – later shown to be wrong – had excluded their existence. The German physicist Dieter Haidt, one of the discoverers, underlined well this absence of interest:

> *During the two-day meeting in November 1968 at Milan, where the Gargamelle collaboration discussed the future neutrino program, the expression 'neutral current' was not even pronounced and, ironically, as seen from today, the search for neutral currents was an also-ran, low in the priority list and subsequently appearing in the neutrino proposal at place 8.* (Haidt 2005)

The situation changed in the following years with the proposal of the unification of the electric and weak forces – in particular by Sheldon Glashow, Abdus Salam and Steven Weinberg (Fig. 4.16). The unification was based on the existence of the mediator Z^0 and, as a consequence, of weak neutral currents.

At the end of the 1960s this new perspective gave an enormous boost to the experimental groups which were searching for this new phenomenon. Finally the experimental proof was announced in 1973, when the Gargamelle bubble chamber – after having been traversed by millions of billions of neutrinos – photographed three events like the one in Fig. 4.17; a neutrino hits an electron from an atom of the heavy liquid, sets it in motion and changes direction.

The discovery was possible thanks to the invention of the 'van der Meer horn', a special magnetic lens designed by the Dutch engineer Simon van der Meer. This device focused pions produced with the Proton Synchrotron so that, as large numbers of them decayed in flight, they would emit a much greater flux of neutrinos traversing Gargamelle than what could be obtained with the Pontecorvo-Schwarz method.

The event of Fig. 4.17 is explained by a diagram similar to the one of Fig. 4.15: the interaction of the neutrino with the electron is mediated by the exchange of the neutral Z^0 asthenon.

The discovery of neutral currents was announced by the Gargamelle collaboration in summer 1973, where it became the hot topic of the Aix-en-Provence Conference.

---

[12] The more than 50 physicists who signed the discovery paper came from – as well as the Orsay Laboratory in France and from CERN – the Paris École Polytechnique, University College London and the universities of Aachen, Brussels and Milan.

Fig. 4.16 Sheldon Glashow (*left*), Abdus Salam and Steven Weinberg at the 1979 Nobel Prize ceremony. Their unification of the electric and weak forces was recognized by the Nobel Committee 4 years before the observation of the mediators of the weak force by Rubbia and collaborators (© Bettmann/CORBIS)

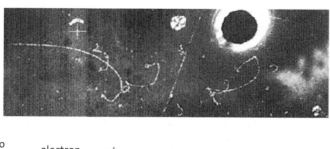

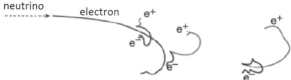

Fig. 4.17 A high-energy neutrino (more precisely, a muon-antineutrino) encounters an atomic electron and sets it moving. The fast electron leaves a track curved by the magnetic field of the chamber and then, colliding with some atomic nuclei, radiates three photons which in their turn produce three characteristic electron-positron pairs (Courtesy CERN)

In his Nobel lecture Abdus Salam recounted the following anecdote: "*I still remember Paul Matthews and I getting off the train in Aix-en-Provence and foolishly deciding to walk with our rather heavy luggage to the student hostel*

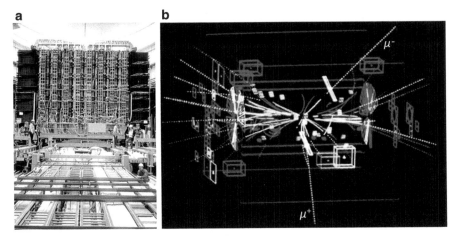

Fig. 4.18 In the detector UA1 (Underground Area 1) (**a**), constructed under the leadership of Carlo Rubbia, the production of Z^0s was observed by detecting – in a large system of 'wire chambers' – their decay into a muon-antimuon pair, like the one shown in figure (**b**), or into an electron-positron pair (Courtesy CERN)

where we were billeted. A car drove from behind us, stopped, and the driver leaned out. This was Paul Musset [of the Gargamelle collaboration] whom I did not know well personally then. 'Are you Salam? Get into the car. I have news for you. We have found neutral currents.' I will not say whether I was more relieved for being given a lift because of our heavy luggage or for the discovery of neutral currents. At the Aix-en-Provence meeting, that great and modest man, Lagarrigue, was also present and the atmosphere was that of a carnival – at least this is how it appeared to me" (Salam 1979).

The events observed with Gargamelle in 1973 constituted certain but *indirect* evidence for the existence of neutral currents, and therefore of the asthenon Z^0; however they did not have the weight of a *direct* proof. In fact, a neutral current event is caused by the exchange of a *virtual* Z^0, while to claim the discovery of the new unstable force-particle one had to observe the production of *real* Z^0's and their subsequent decays into other stable particles.

The discovery of real Z^0's arrived 10 years later, when Carlo Rubbia (Fig. 3.7 on p. 74) and his collaborators observed the production of the asthenons W^+, W^- and Z^0. The experiment was carried out with the CERN proton-antiproton collider, thanks to the application of stochastic cooling invented by the same extraordinary van der Meer, as was already discussed in Chap. 3 (see Fig. 3.6 on p. 73).

Carlo Rubbia not only proposed and directed the transformation of the Super Proton Synchrotron (SPS) at CERN into a proton-antiproton collider, but also led the design and construction of the huge particle detector which observed the decays of scores of the W and Z asthenons (Fig. 4.18).

By measuring the energies and directions of the decay products, the masses of the three force-particles were also determined to be around 80 GeV for the charged asthenons W^-, W^+ and about 90 GeV for the neutral asthenon Z^0.

Chapter 5
In Search of the Higgs Field

Contents

© Springer International Publishing Switzerland 2015
U. Amaldi, *Particle Accelerators: From Big Bang Physics to Hadron Therapy*,
DOI 10.1007/978-3-319-08870-9_5

Daniele Bergesio, TERA Foundation

Many meetings are held in the Georges Charpak Conference Room, *on the sixth floor of the CERN Main Building, among which the annual sessions of the Scientific Policy Committee are particularly important. The room is named after an extremely talented experimental physicist who died in 2010. Charpak made crucial contributions to the invention and development of particle detectors, without which it would be useless to construct increasingly powerful accelerators.*

An extraordinary, dynamic and charming personality, Georges Charpak was born in 1924 in the ghetto of Sarny, Poland, from where he moved with his parents to Palestine, and finally to France, his adopted country, at the age of eight. Outstanding at school, he was able to skip two years at the prestigious Lycée Saint-Louis in Paris. In 1943 he was arrested for Resistance activities and deported to Dachau, which he survived by succeeding in hiding his Jewish origin thanks to his knowledge of languages and the brilliant blue colour of his eyes.

At the end of the war he enrolled as a foreign student at the École des Mines in Paris, where he graduated three years later as an engineer, but very soon became fascinated by physics working in the laboratory of the husband and wife team, and Nobel laureates, Jean-Frédéric Joliot and Irène Curie. In later life, he was to write: "Physics is the most demanding and, sometimes, the most destructive of lovers. Night and day, summer and winter, morning and night it persecutes you, invades you, fills you with satisfaction and despair" (Charpak and Saudinos 1993).

Charpak worked at CERN from 1959 when, during the mid-1960s, he became convinced that, to record and study not just a few but many thousands of collision events per second, new types of detectors were essential. Bubble chambers were too slow, because they required the movement of a large piston to bring the liquid to the ideal conditions of pressure and temperature.

Thus it was that Georges Charpak invented and built, with the aid of his collaborators, the first 'multiwire chamber' with dimensions ten centimetres on each side. It was constructed as an array of about thirty parallel metal wires, each as fine as a human hair, immersed in a special gas mixture. During the passage of a charged particle, the gas became ionised and small clouds of electrons produced along its path were rapidly collected by the chamber wires; the arrival of the particle was thus converted into a fast electrical pulse which, transmitted along the wires, could eventually be registered and stored by computers, which were becoming more widely used at that time.

Over a few years, wire chambers of increasing size and greater spatial precision, like 'drift chambers', were constructed and proved to be essential for research in colliding beam experiments. Today the giant detectors assembled at the Large Hadron Collider are equipped with wire chambers, which cover areas equivalent to many hectares and are capable of recording the interesting events among the billions of proton-proton collisions taking place each second.

Having received the Nobel Prize for physics in 1992, Georges Charpak continued to invent new particle detectors and apply them to medical diagnosis. A notable public and political ambassador for CERN activities, with his contagious enthusiasm he conveyed his passion for science to people, who either followed his conferences or met him in person, and in interviews with newspaper and television journalists. In France he became a public figure known to all as the principal promoter of a popular science teaching programme, "Getting your hands dirty", for French elementary school children, which is still much followed and, since 2012, is coordinated by "La main à la pâte Foundation".

The Standard Model of Forces and Particles

What physicists call 'The Standard Model' assembles all the properties of the subatomic world into a coherent theoretical picture; up to now its predictions have been confirmed by thousands of experiments. The Model is based on the *24 matter-particles* of Fig. 4.6 on p. 105, on *quantum chromodynamics* (which describes the strong force that acts on coloured quarks) and on an elegant description which *unifies* the electromagnetic force (which acts on all charged particles) with the weak force (to which all particles are subject). This 'electro-weak unification' is discussed later in this chapter.

The 30 force-particles, mediators of the four fundamental interactions, are listed in Table 5.1, where the final column records the approximate value of each coupling.

The eight gluons, whose coupling is of order unity, mediate the 'strong' force. The photons and the two asthenons have couplings of the order of 1 %. The 30th mediator is the hypothetical 'graviton', which has an infinitesimal coupling to matter-particles and whose effects are negligible in the subatomic world; for this reason the Standard Model does not take into account the exchange of gravitons.

The photon is massless; therefore the range of the electric force produced – determined by the relation $R = 0.2/M$ – is infinite. In practice this means that the force is felt even at large distances from the matter-particle, although very feebly. In contrast, the asthenons W and Z have very large masses, 80.4 GeV and 91.2 GeV, and make themselves felt only over distances less than two thousandths of a fermi: $0.2/100 = 0.002$ fm.

Figure 5.1 illustrates the complete Standard Model, which contains both the 24 matter-particles as well as the 12 force-particles which act in the subatomic world; 8 gluons, 1 photon and 3 asthenons.

Table 5.1 The mediators of the four fundamental forces act in different ways on the matter-particles

Force	Mediators	Symbols	Mediator mass in GeV	Matter-particles on which the force acts	Coupling
strong	8 gluons	$g_1; g_2; g_3; g_4;$ $g_5; g_6; g_7; g_8$	0	Quarks	about 1
Electric	1 photon	γ	0	Quarks; charged leptons	0.0073
Weak	3 asthenons	$W^+, W^-; Z^0$	**80.4, 80.4; 91.2**	Quarks; charged leptons; neutrinos	0.016; 0.010[(*)]
Gravitational	1 graviton	G	0	Quarks; charged leptons; neutrinos	10^{-45}[(**)]

[(*)]The W and Z couplings depend on the pair of matter-particles with which the mediator exchanges energy. The values in the table refer to electron-neutrino and neutrino-neutrino pairs
[(**)]This value refers to the gravitational force between two electrons

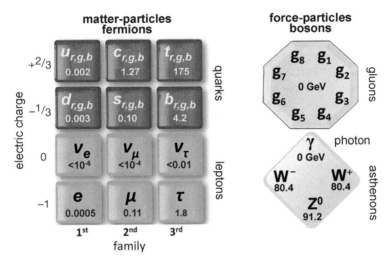

Fig. 5.1 A graphical representation of the Standard Model of matter-particles (which are 'fermions') and force-particles (which are 'bosons')

Matter-particles and the force-particles differ in a very important characteristic. To illustrate it let us recall that with light it is possible to produce a laser beam made up of billions upon billions of photons which all move simultaneously in the same direction, without diverging. This magnificent technological achievement is possible because photons are boson*s*, which is to say particles that 'like' – speaking anthropomorphically – to group together in the same state of motion in great numbers.

Electrons, which are matter-particles, behave in exactly the opposite manner. When they circulate around an atomic nucleus, each one moves in its own unique orbit, avoiding occupying one already taken by another electron. Such particles, which cannot be made to share their own state with other identical particles, are called fermions.

As indicated in Fig. 5.1, all the force-particles are bosons and all the matter-particles are fermions.[1] Therefore, often 'boson' is used synonymously for 'force-particle' and 'fermion' as a synonym for 'matter-particle', and one says that *a force is the consequence of the exchange of a virtual boson between two fermions.*

---

[1] The word *boson* originates from Savendra Nath Bose, an Indian physicist who in 1924 wrote to Einstein to discuss the special properties of photons. Instead *fermions* take their name from Enrico Fermi who, at the age of 24, published his theory describing the behaviour of electrons, so becoming notable as one of the great physicists of the era.

The LEP Electron-Positron Collider

The construction of electron-positron collider LEP was led by the physicist Emilio Picasso, chosen by Herwig Schopper who had been nominated CERN Director-General from 1 January 1981. As Schopper explained 30 years later, the choice created not a little discontent among the accelerator experts, while it was much welcomed by all the experimental physicists.

> Among the accelerator people there were some who had been responsible for earlier projects, had played a decisive role in the preparation of the earlier LEP proposal and had all the qualities to promise full success for the LEP project. Hence, many were quite surprised or even disappointed when I asked the Italian Emilio Picasso to lead the LEP project. He was an experimental high-energy physicist and not an accelerator expert. However, during his career he did experiments which required close collaboration with the technical divisions and therefore he enjoyed enormous respect and recognition not only among the physicists but also among the accelerator engineers. In addition, he had the temperament and the personality to smooth out human problems. It turned out that this was an excellent decision. Emilio Picasso did a superb job. (Schopper 2009)

When Schopper and Picasso took up their posts, the LEP design foresaw a 30 km ring located about 100 m below the plains which lie between Geneva and the French Jura mountains, in the middle of which is located Geneva airport. The choice of the position of the tunnel was a very difficult decision which took more than a year to make. On one hand it was desirable to dig the longest possible tunnel to reach the highest possible energies, both for LEP as well as for a possible future LHC, which was then already foreseen to be built using the same tunnel. On the other hand, the terrain between the airport and the foot of the Jura – a special kind of sandstone called 'molasse' – was ideal for tunnel excavation, while the limestone under the Jura was very bad. Schopper has described this period vividly:

> The real great danger of the initial design was the 12 km tunnel under the Jura with rock of calcareous nature with big caverns and possible infiltrations of water under high pressure. After long and sometimes heated discussions we decided to reduce the circumference of LEP from 30 km to 27 km and move it somewhat out of the Jura to reduce the geological risks, but 8 km of tunnel were still under the Jura.
>
>   When I presented the proposal for LEP to the CERN Council of June 1981 this position seemed a reasonable compromise. However, on 9 October 1981 Picasso sent me a confidential memorandum in Italian, the language he used when he was very worried. In the memorandum he summarized the evaluation of the geological risks based mainly on the final advice of an outstanding geology expert, Professor Giovanni Lombardi from ETH Zürich.
>
>   All his arguments led to a hazardous and gloomy prospect for the tunnelling and the whole project; hence, we decided to displace the main ring towards the east by several hundred metres, bringing it more out of the Jura and closer to the airport. This position offered several decisive advantages, the main one being that the length of the tunnel under the Jura was reduced from 8 to 3 km. Of course in real life it is not possible to gain only advantages; the geological surveys had shown that the roof of the molasse was descending towards the airport. To keep the tunnel in the molasse, it had to be rather deep near the airport. On the other hand, we wanted it to be as close to the surface as possible at the foot of the Jura. These two conditions could be fulfilled only by putting the tunnel on an inclined plane with a slope of 1.4 %. LEP was the first circular accelerator installed on an inclined plane. (Schopper 2009)

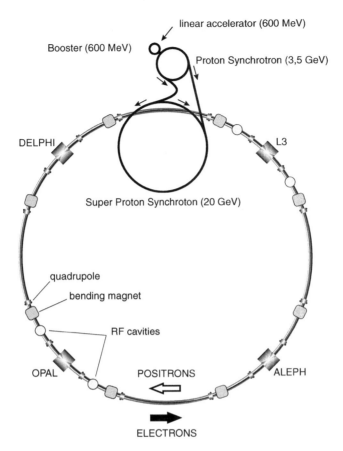

Fig. 5.2 A schematic plan view of LEP, 27 km in circumference installed in a tunnel excavated 100 m underground, now occupied by the LHC

In spite of all the precautions, in September 1986 water with a pressure of 12 atm broke into the tunnel from a geological fault. Eventually the water flow could be reduced and the tunnel drained, with its roof and sidewalls supported by iron and concrete vaults. This caused an 8 month delay in the construction, so that LEP commissioning was completed in July 1989.

Between 1989 and 2000 at LEP (Figs. 3.5 on p. 71 and 5.2) beams of electrons and positrons circulated in opposite directions in a single 27 km long aluminium vacuum pipe, which had a cross-sectional diameter of a few centimetres.

Starting from a linear accelerator and the PS, the original ancestor of all CERN accelerators, and passing via the Super Proton Synchrotron, electrons and positrons were injected into the LEP beam pipe with energy of 20 GeV. Once synchronised to be in phase with the 'circulating beams', the kinetic energy of the particles was increased using radiofrequency cavities, placed at intervals along the ring to accelerate the particles by means of alternating electric fields.

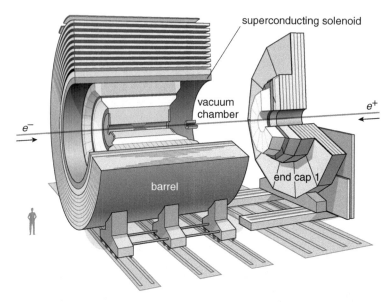

Fig. 5.3 DELPHI, like almost all collider detectors, was constructed as a 'barrel' with two 'endcaps', which were closed during experimental operation but movable on sliding rails to allow access to the interior sub-detectors

LEP Particle Detectors

Having reached a chosen energy in the range 20–100 GeV, the electrons and positrons were allowed to circulate for 1 or 2 days to record annihilations which took place at each of the collision points where one of four large detectors, L3, ALEPH, OPAL and DELPHI, were located. In almost 10 years, each experiment recorded more than five million electron-positron annihilation events.

It is natural for me to take the example of the DELPHI detector of Fig. 5.3, since I was its scientific leader ('spokesperson') from the original conception in 1980, until 1993.

DELPHI (*Detector with Lepton, Photon and Hadron Identification*) was an assembly of detectors 12 m high and 15 m long, designed and constructed in 9 years by a collaboration of about 500 physicists originating in 20 different countries.

As in all the large detectors installed at particle colliders, the 'barrel' of Fig. 5.3 was composed of many concentric 'sub-detectors', each constructed by a collaboration of different universities and laboratories and then transported to CERN for final assembly. Compared to the other LEP detectors, DELPHI had a particular objective of measuring hadrons with special precision, permitting their identification even when they disappeared after only a millionth of a microsecond.

Fig. 5.4 This picture shows (from *left* to *right*) Georges Charpak with his collaborators Fabio Sauli and Jean-Claude Santiard working on an ISR wire chamber (Courtesy CERN)

The sub-detectors – constructed in several different ways to be well adapted to measure the passage of different types of particle – were mainly based on wire chambers, which covered thousands of square meters. Twenty years had passed since Charpak's invention and physicists had learned how to build larger and larger wire chambers. Figure 5.4 shows the first really large chamber that Georges Charpak built with his collaborators in the 1970s to detect the particles produced in proton-proton collisions at the CERN Intersecting Storage Ring (ISR).

In colliders, such as the ISR and LEP, the newly created particles emerge from the collision point at almost the speed of light and deposit tiny electronic signals in the different layers of the detector. These signals are amplified, recorded and displayed on the screens of powerful computers as continuous tracks (Fig. 5.5).

The recorded signals are like a signature that the particle engraves on the detector; however this signature can be interpreted only by specific computer software, which permits identification and measurement of the most important properties of the particle, which are the energy and direction of motion.

For a charged particle the energy can be obtained from the curvature of the trajectory; the more curved, the lower the energy. The bending is due to the strong magnetic field created by the giant coil of the superconducting solenoid in Fig. 5.3, which at the time was the largest ever built. The cylindrical coil was wound using a

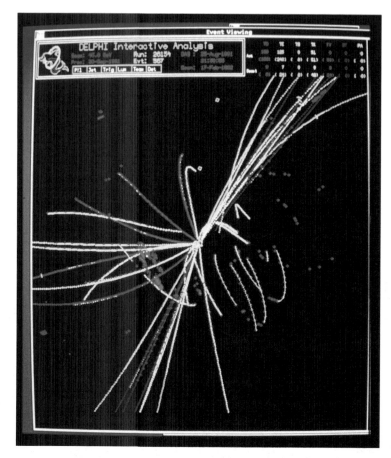

Fig. 5.5 This event recorded at LEP by DELPHI shows the tracks – bent by the magnetic field – *left* by hadrons and leptons created in an electron-positron annihilation at 90 GeV (CERN/Science Photo Library)

cable made of a special metal alloy which, when held at 270° below zero in a bath of liquid helium, has zero electrical resistance; it therefore dissipated no energy despite carrying a current of many thousands of amperes.

Fundamental Fields rather than Particles

In a typical annihilation event an electron and positron disappear, to be replaced by dozens of particles that have nothing to do with the original electron and positron. It is as if the disappearance of two strawberries gave rise to a flood of nuts, bananas, pears, apples and fruits of every kind (Fig. 5.6) – and to their 'antifruits', which are not shown in the figure.

Fig. 5.6 In the annihilation of two particles (strawberries) *different* types of particles (fruits) appear. How is that possible? The answer is given by quantum field theory

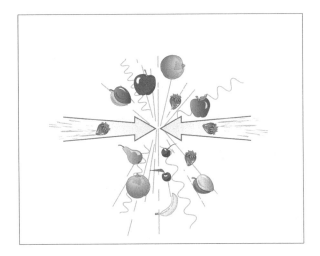

How do the strawberries that collide 'know' that there exist apples, bananas and pears, so as to give way, when disappearing and releasing energy in the collision, to fruits *different* from strawberries?

Beyond the metaphor, when the electron and positron collide in vacuum, annihilating as in Fig. 5.5, where 'is it written' that 18 types of quark-antiquarks and 3 types of lepton-antilepton pairs exist, into which the flash of pure energy produced in the annihilation can materialise? The information cannot be contained in the colliding particles, the electron and positron; they are point-like bodies which, so to speak, "know nothing beyond themselves".

The solution found by physicists to this mystery is that empty space, in which the annihilation and the creation of new particles takes place, cannot be truly *empty*; it must actually be occupied by ethereal entities – which physicists call 'fields' – that contain the necessary information.

In the absence of energy, all the fields are quiescent and appear inactive. But when energy is injected, one of the fields becomes excited in the form of a *local perturbation*, a 'ripple' which can propagate through empty space carrying away some of the energy.

It resembles what happens when a sudden gust of wind creates a wave on the surface of a calm lake, which then spreads outwards from the original point of its creation. The lake already existed, like the fields in space, but the energy from the breeze produces a limited but very identifiable perturbation, which propagates along the surface of the water.

The Japanese theoretician Kazuhito Nishijima gave an original explanation for introducing fields in the description of the creation and annihilation of particles and antiparticles.[2]

---

[2] In 1953 Nishijima introduced an equation that relates various fundamental properties of each hadron. Discovered independently 3 years later by Murray Gell-Mann, it is now understood to be a direct consequence of the quark structure of all hadrons.

The situation may be compared to the police activity of finding a criminal. The most natural way is to let detectives follow and watch all the suspects until they discover some evidence. However, if a suspect suddenly escapes from the observation of the detective following him, the investigation becomes incomplete and the police might fail to find a clue. Instead, one can think of a better, more complete way of finding the offender. Namely, send as many detectives as possible to all important spots and let every detective report what happened at every spot. In this way every spot can be covered and one can expect complete observation of the suspects. In the world of fundamental particles, similar things happen (in particular, a particle-antiparticle pair can either appear or disappear) and the second method of observation is needed.

It is an especially convenient method of description to give the number of various particles in every small space domain. Then it is possible to describe phenomena in which creation and annihilation of particles take place, and this is essentially the field theoretical description. (Nishijima 1965)

According to this picture, every fundamental particle is nothing else but a localised perturbation in the corresponding fundamental field; it is a 'quantum' which travels from one point in space to another, carrying away a well defined quantity of mass and energy. So, for example, a photon is a quantum of the *electromagnetic field*, a gluon is a quantum of one of the 8 *gluon-fields*, an electron is a quantum of the *electronic field*, a *u*-quark is a quantum of the *u-quarkonic field*, and so on.

Visualising Particles as Waves

The fundamental fields extend throughout the whole of three-dimensional space. But to visualise a particle as a localised perturbation in a field, we should imagine just a *two-dimensional* world, as in Fig. 5.7; here the objects are *flat* and move in a plane without ever being able to exit from it, and the observers are like *flat* ants which can perceive only what happens on the surface of the plane, without any means of seeing what happens above or below it.

Here an inactive field is represented by a grid of parallel lines which lie in the horizontal plane where the phenomena occur. The oscillations of each field are described by the deviation of the lines from the quiescent position. Figure 5.7 shows three successive states of the two different fields, with the electromagnetic field represented by red lines and the *u*-quarkonic field by black lines orthogonal to the red ones.

Initially the *u*-quarkonic field is in a quiescent state while the electromagnetic field exhibits a local oscillation that corresponds, for example, to a virtual photon created by the annihilation of the 25 GeV electron with an equal energy positron. After a very short delay, defined by the uncertainty principle, this photonic oscillation disappears giving all its energy to the *u*-quarkonic field. Two local u and $\bar{u}$ black perturbations then appear in the quarkonic field, which move apart, each carrying off 25 GeV of energy.

Fig. 5.7 The annihilation of an electron and a positron excites the electromagnetic field; the virtual γ disappears instantly, injecting all its energy into the quarkonic field and producing two waves, one in the form of a peak (the *u*-quark) and the other in the form of a trough (the *ū*-antiquark) (Paolo Magagnin, TERA Foundation)

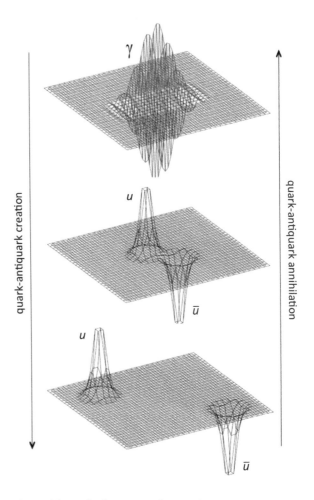

quark-antiquark creation

quark-antiquark annihilation

γ

u

ū

u

ū

Alternatively, if the figure is read from the bottom to the top, in reverse, one sees a peak filling a trough where the *u*-quark annihilates the *ū*-antiquark; the black quarkonic field de-excites while the red electromagnetic field becomes excited.

The ants which live in the two-dimensional plane do not see the peaks and troughs which form the ripples in the field; they see only the local perturbations which appear in the plane and call them 'particles'. Only those who are privileged, by being aware of the third dimension, can see that what appears to be a particle is the undulation of a pre-existing field, extending over a small region of space.

From this new point of view, *the particles are nothing other than localised waves*; therefore I like to call them *wavicles*, a term that combines the nouns 'wave' and 'particle'.[3] The wavicles of the electromagnetic and asthenonic fields are photons and asthenons; the wavicles of the u-quarkonic field are the u-quarks and the $\bar{u}$-antiquarks, those of the electronic field are the electrons and positrons, and so on.

The fields which behave in this way are said to be 'quantum fields' and the theory which describes them quantitatively is known as 'quantum field theory'.

The first theory to be constructed in the 1950s was *quantum electrodynamics,* which explains with great precision phenomena in which photons, electrons and positrons (or equivalently photonic and electronic fields) are involved as a result of the electric coupling. The theory was easily extended to include all charged particles, and has served as a model for constructing quantum field theories of the weak and strong forces. In the case of the strong force, the theory of *quantum chromodynamics*, which has already been mentioned, succeeds in incorporating in a single scheme all experimental results involving quarks and gluons. Its fundamental entities are the 18 quarkonic and 8 gluonic fields.

The descriptions based on quantum fields make sense of the otherwise mysterious *wave-particle dualism* of physical quanta; the particles are waves when they propagate and when they interfere, but appear as corpuscles when they are detected in experiments.

For example, a wavicle gives all its energy to a granule of photographic emulsion, so chemically exciting it, and its arrival point is revealed when, following the chemical development process, these tiny grains blacken. It is the wave that determines the probability of finding the particle at a particular point when it is observed; this justifies the fact that the term 'wavicle' reminds one more of the wave than the particle.

In quantum field theories the *fundamental entities* are the fields, impalpable media filling all space, and not the particles and the antiparticles. From this perspective the Standard Model is based, not on 36 particles but on 36 quantum fields, the origin of 24 types of matter-waves and 12 types of force-waves. However, the depiction of Fig. 5.1 (p. 125) is still valid, because the symbols that accompany it can be interpreted as representative of the fundamental fields, instead of the particles.

Particles and antiparticles are nothing more than ephemeral manifestations of the field, undulations which appear and disappear easily. Only the low mass wavicles, that is electrons and u- and d-quarks, are exceptions; they cannot disappear because there are no wavicles of lower mass into which they can decay; this is why they are unique in being the stable constituents of nuclei and atoms.

---

[3] This neologism was introduced in 1927 by the great British physicist and astronomer Arthur Eddington, who in 1919 travelled to the island of Príncipe near Africa to watch a solar eclipse and measured the deflection of the star light rays close to the Sun, a phenomenon predicted by Einstein's theory of general relativity according to which the masses bend the space around them.

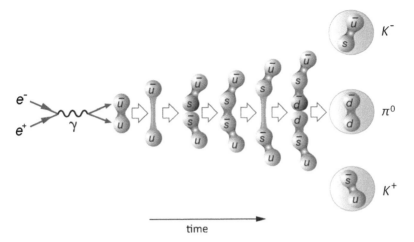

time

Fig. 5.8 The 'hadronisation' phenomenon is here described with the simple language of particles and antiparticles instead of the more complicated field language: as soon as they form, $s\bar{s}$ and $d\bar{d}$ pairs – as the earlier quark-antiquark pairs – separate two by two, combining into the final mesons (here two oppositely charged kaons and a neutral pion)

Isolated Quarks Cannot Be Observed

At LEP, which had a maximum energy of 200 GeV, almost all the matter-particles described by the Standard Model of Fig. 5.1 were produced; or better, in the parlance of quantum fields, almost all fields were *excited*. In particular in 70 % of electron-positron annihilations a large amount of energy was released in the form of excitations of the quarkonic fields, i.e. as quark-antiquark pairs; for example a u-quark and its antiquark or a s-quark and its s-antiquark, which has a different flavour. All the quark-antiquark flavours were produced except the t-quark and its antiquark, because to produce them would have required an energy at least equal to their total mass, namely 2×175 GeV $= 350$ GeV.

However, quarks and antiquarks cannot just freely escape. Figure 5.8 illustrates the phenomenon of 'hadronisation', i.e. the sequence of events initiated by a newly produced quark-antiquark pair, which ends with the production of many hadrons.

The quark and the antiquark, created by the flash of energy liberated in the electron-positron annihilation, move apart in opposite directions, but not so that they can be observed as free particles, because the 'gluon elastic' holds them tightly bound together. However, at a certain point the elastic breaks with the production of a new quark-antiquark pair. The phenomenon repeats until all the initial energy is transformed into the mass of a large number of quarks and antiquarks of different flavours, which ultimately are bound together in mesons by the strong force.

Sheldon Glashow (Fig. 4.16 on p. 118) convincingly described a similar phenomenon: the extraction of a quark from the proton in which it is bound by the strong force.

> *The electromagnetic force between two charged particles is described by Coulomb's law: the force decreases as the square of the distance between the charges. The strong force between two coloured quarks behaves quite differently: it does not diminish with distance but remains constant, independent of the separation of the quarks, and an enormous amount of energy would be required to isolate a quark. Separating an electron from an atom requires a few electron volts. Splitting an atomic nucleus requires a few million electron volts. In contrast with these values, the separation of a single quark by just an inch from the proton of which it is the constituent would require enough energy to raise the author from the earth by some 30 feet.*
>
> *Long before such an energy level could be attained another process would intervene. From the energy supplied in the effort to extract a single quark, a new quark and antiquark would materialize. The new quark would replace the one removed from the proton and would reconstitute that particle. The new antiquark would adhere to the dislodged quark, making a meson. Instead of isolating a coloured quark, all that is accomplished is the creation of a colourless ('white') meson.* (Glashow 1975)

The German physicist Ralph Shutt, who contributed to the discovery of the Ω^- (Fig. 4.3 on p. 98), introduced another forceful analogy. A quark bound in a hadron "*is reminiscent of an iceberg, nine-tenths of which is under water and not directly visible: it is bound to the earth by gravity. To see the whole iceberg and to explore its properties, unencumbered by the presence of the ocean water, one must lift it out of the water, requiring a large amount of energy*" (Schutt 1971). In the case of quarks, because of the gluon elastic, the amount of energy is infinite and quarks *cannot* be liberated at all.

Glashow also made an interesting general remark on the long quest to discover, with particle accelerators, the most intimate components of the fundamental particles:

> *If this interpretation of 'particle confinement' is correct, it suggests an ingenious way to terminate the apparently infinite regression of finer structure in matter. Atoms can be analysed into electrons and nuclei, nuclei into protons and neutrons, and protons and neutrons into quarks, but the theory of quark confinement suggests that the series stops here. It is difficult to imagine how a quark could have an internal structure if it cannot even be created.* (Glashow 1975)

Similarities Between Gluons and Photons

In 30 % of all high-energy electron-positron annihilations, lepton-antilepton pairs are produced. In the remaining 70 %, more complicated processes than the ones represented in Fig. 5.8 take place. The hadronisation of the quark and the antiquark creates about *forty* new quark-antiquark pairs. They rearrange themselves in innumerable ways, giving rise in the final state to, on average, 20 hadrons, including – although much less often – baryons and antibaryons, which are quark and antiquark triplets. This is schematically depicted in Fig. 5.9.

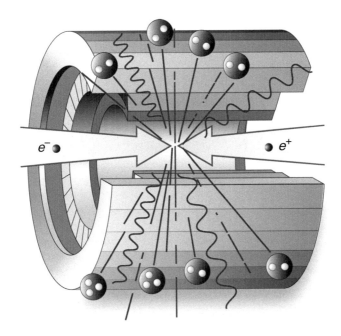

e^- ●   ● e^+

Fig. 5.9 In LEP annihilations dozens of bound states made of three quarks or antiquarks (baryons or antibaryons) or of a quark and antiquark (mesons) were produced

This figure illustrates the general phenomenon, but not its most important characteristic; almost all the final state hadrons – created by the hadronisation of a quark or an antiquark – move in the same direction as the initial hadron which produced them and therefore appear in the detectors as collimated 'jets' of almost parallel tracks.

One of these jets, which is made essentially of hadrons and is called a 'quark-jet', appears clearly in the upper part of Fig. 5.5. Instead, in the lower part of the figure not one but two jets can be seen. The event is represented schematically in Fig. 5.10a: one of these jets was caused by the antiquark, while the other is due to the emission of a gluon from the quark. Overall the antiquark created from the flash of energy gives rise – through the hadronisation process – to the antiquark jet which goes upward, while the quark emits a virtual gluon and subsequently produces the 'quark-jet' which is directed downwards.

The emitted gluon is 'coloured' and, not being able to exist as a free particle, produces in its turn a 'gluon-jet' following the hadronisation phenomenon analogous to the one undergone by the quark and antiquark.

Therefore, a large number of mesons (particles made of a quark and antiquark), and some baryons and antibaryons (made of three quarks or three antiquarks) emerge from the point where the electron-positron annihilation took place, but in no case does a quark, antiquark or gluon appear. However, the existence of these fundamental particles subject to the strong force is proven experimentally by the fact that the final hadrons do not emerge with equal probability in all directions, but form collimated

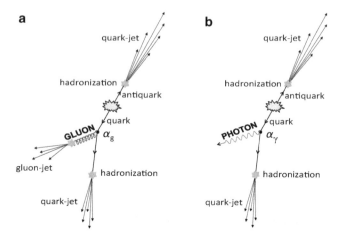

Fig. 5.10 (a) The creation of a quark, an antiquark and a gluon appear in the detector – following the three phenomena of hadronisation – in the form of a three-jet event. (b) A similar event occurs with the creation, due to electromagnetic force, of a photon instead of a gluon; the photon travels very far, while the gluon does not exist as free particle and immediately transforms into a jet of hadrons, i.e. in a 'gluon-jet'

jets of particles. These jets have their origin in the large energy of the initial quark or antiquark, which in the hadronisation process is shared among a group of hadrons whose trajectories retain the memory of the direction of their parent particle.

For this reason physicists classify the events recorded in their detectors by counting the number of jets; two jets indicate the creation of a quark-antiquark pair, three jets signal that one of these particles has emitted a gluon. The experimental proof of the existence of the gluon was in fact obtained in 1979, at the DESY (Deutsches Elektronen-Synchrotron) laboratory in Hamburg, by observing a few 'three-jet events'.

The electron-positron collider PETRA (Positron-Elektron-Tandem-Ring-Anlage) had been operating for a year when, in the summer of 1979, it reached 27.4 GeV in energy. A week later Bjørn Wiik – spokesman of the TASSO collaboration (Fig. 3.14 of p. 83) – presented the event of Fig. 5.11a, which has an evident three jet structure, at a conference in Bergen (in Norway, his native country). In the succeeding weeks further similar events were observed by TASSO and by the other collaborations (JADE, MARK-J and PLUTO) which had installed their detectors at the PETRA collision points. The discovery of the gluon at DESY was the first of a series of experiments which confirmed the validity of quantum chromodynamics with increasing precision.

The observation of three jet events made at PETRA demonstrated experimentally that a quark – which carries a colour charge – sometimes emits a 'massless gluon' through the process of Fig. 5.10a, which is analogous to the emission of a 'massless photon' by the quark of Fig. 5.10b. However, this latter phenomenon occurs a 100 times less often because the electric coupling α_γ is 100 times smaller than the strong coupling α_g.

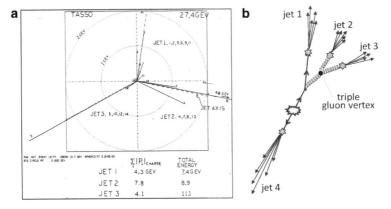

Fig. 5.11 (**a**) This three-jet event – presented by the TASSO collaboration at a conference in Bergen in 1979 – was the first experimental observation of the existence of the gluon, mediator of the strong force (Courtesy of Sau Lan Wu). (**b**) The diagram describes a typical four-jet event used for the measurement of the triple gluon vertex, which was presented by the DELPHI collaboration at the 1990 Singapore conference

Differences Between Gluons and Photons

The similarity between the strong and electric forces is evident from comparison of the diagrams of Fig. 5.10a, b and from the fact that the nine force-particles, mediators of the two forces, are all massless. For this reason quantum chromodynamics was constructed following the example of quantum electrodynamics.

However, comparing chromodynamics with electrodynamics, it is clear that *gluons are both very similar but also very different* from photons. The reason of the difference between photons and gluons, in spite of their significant similarities, is a profound one and is related to the very existence of the forces which act in the subatomic world and their unification, today's frontier in the physics of high energies about which more will be said in the next chapter.

We start from the observation that, in the world around us, matter and forces are different aspects of the same reality; matter is mass and the forces push and pull on it. In the 1930s the description of the subatomic world, based on the uncertainty principle, permitted the unification of these two concepts making it evident that forces are due to the exchange of specific types of particle, the force-particles. However – in the Standard Model of Fig. 5.1 (p. 125) – the matter-particles (which carry mass) and the force-particles (which cause the forces) still have completely different natures, beginning from the fact that the first are fermions and the second are bosons.

In the 1950s it was understood that the electric force is a *necessary* consequence of the existence of the electronic field, once it is required that such a field has a particular form of symmetry which was called 'gauge symmetry'.

It is not easy to explain in simple terms what this symmetry is and how it *imposes* the existence of the photonic field and therefore the electric force, transmitted by the exchange of photons. Nevertheless the argument is important – because it also applies to the strong and weak forces and is at the foundation of the unification of all the forces, so it is worth the trouble to try to understand it, at least qualitatively, reading (and rereading. . .) some lines of Steven Weinberg (Fig. 4.16 on p. 118):

> *The good news is that there is a class of quantum field theories, known as 'gauge field theories' which offer the prospects of unifying weak, electric and strong interactions in one elegant mathematical formalism. To explain these theories, I should like to make reference to something more familiar.*
>
> *One of the fundamental principles of physics is that its laws are not dependent on the orientation of the laboratory. But suppose one were to establish a rotating laboratory in which the orientation would change with time. Would the laws of nature then be the same? Newton (with his mechanics) would say no; he would maintain that the rotation of a laboratory with respect to absolute space produces a change in the form of the laws of nature. This change can be seen, for example, in the existence of centrifugal forces.*
>
> *In his theory of General Relativity[4] Einstein, however, gives a different answer; he says that the laws of nature are exactly the same in the rotating laboratory as in the laboratory at rest. In the rotating laboratory, the observer sees an enormous firmament of galaxies, all the matter of the universe, rotating in the opposite direction. This flow of matter and energy produces a field which acts on the laboratory and, in turn, produces observable effects like centrifugal force. In Einstein's theory, that field is gravitation. In other words, the principle of invariance, the ideas that the laws of nature are the same whether in a rotating laboratory or one at rest, requires the existence of the force of gravitation. Moreover, this gravitational field is responsible not only for centrifugal forces but also for the ordinary gravitational force between the Earth and Newton's apple.*
>
> *Gauge theories are similar; they are theories in which there is a principle of invariance which logically requires the existence of the forces themselves. However, in gauge theories, the principle of invariance or symmetry – I use the words interchangeably – is not the spatial symmetry with which we are familiar but an internal symmetry. Whenever such symmetries arise, they force the matter-particles to fall into neat families – doublets, triplets, etc. The force-particles form a family themselves whose membership is determined by the nature of the symmetry.* (Weinberg 1977)

The gauge symmetry is not a symmetry of space but a symmetry of the field which describes matter-particles. In the case of the electronic field, this symmetry property implies the existence of a force field which can be identified as the 'photonic' field (usually called the 'electromagnetic' field) and has just *one* associated force-particle, the photon. Similarly, from the requirement that the three quark fields u_r, u_g and u_b (and all the other triplets of quark fields) are subject to an analogous form of gauge symmetry, it follows that *eight* force fields should exist, whose quanta are the familiar eight gluons.

Applying the gauge symmetry principle to the electronic and quarkonic fields leads thus to both the electric force and the strong force: the existence of nine force-particles (one photon and eight gluons) follows from the two gauge symmetries and need not to be introduced as independent hypotheses.[5]

---

[4] Some of the consequences of Einstein's 'special' relativity, formulated in 1905, have been discussed at the end of Chap. 1. 'General'' relativity, conceived ten years later, extended special relativity by arguing that inertial forces – i.e. forces due to the inertia of the bodies - are indistinguishable from gravitational forces.

[5] In mathematical language the two *internal* symmetries are identified by the symbols U(1) and SU(3).

But there is more. For reasons it is not possible to explain here, in every gauge field theory the mass of the force-particles should be exactly *zero*. This actually does correspond to photons, gluons and gravitons, mediators of the gravitational force, as indicated in Table 5.1 of p. 124. In this respect photons and gluons appear very similar.

Nevertheless they are also very different, because the strong force produces events similar to the one shown schematically in Fig. 5.11b, which manifests itself as a four-jet event, while events in which the three gluons of this figure are replaced by three photons cannot exist and in fact have *never* been observed.

Four-jet events are characterised by a quantity called the 'triple gluon vertex' – indicated in Fig. 5.11b – which determines the probability of the emission of a gluon by another gluon. It is a very important quantity in quantum chromodynamics, because it implies that gluons interact with themselves forming the 'gluon elastics' depicted in Fig. 5.8 (p. 135); as a consequence the strong forces increases indefinitely when a quark and an antiquark are separated. The 'triple-photon vertex' does not exist and the electromagnetic force decreases with the distance. A measurement of the triple gluon vertex was made in Spring 1990 by the DELPHI collaboration – only 1 year after the first electron-positron annihilations in LEP – and I had the great satisfaction to present it at the international particle physics conference which was held in Singapore in August of the same year. As I will explain in Chap. 8, this event had a major influence on the rest of my professional life.

Electro-Weak Symmetry and the Higgs Field

In Table 5.1 on p. 124 an anomaly is evident: the wavicles of three force-fields – photons, gluons and gravitons – are massless, while the asthenons, mediators of the weak force, are about a 100 times more massive than a proton. The strong, electromagnetic and gravitational forces can be described by gauge field theories and this fully explains the fact that their mediators have no mass. But why this does not happen to the weak force too? Does this mean that the gauge invariance principle does not apply to the weak forces?

Theoretical physicists from the entire world scratched their heads for years over the problem of the asthenon masses and and – in particular – on the difference between the massless photon and the extremely massive Z^0 asthenon, given that their couplings to quarks and leptons are very similar.

After many unsuccessful attempts, a convincing mechanism capable of explaining the mystery was identified in 1964. The starting hypothesis is that the weak force is described, like the others, by a gauge field theory but that, at the same time, the entire universe is occupied by a 37th quantum field that originated in the Big Bang and is known as a *scalar field* or, from the name of one of the theorists who proposed it, the *Higgs field*. The masses of the W and Z asthenons are due to the pervasive presence of this field even if the weak force follows, as do the others, from a gauge symmetry, which automatically implies that their mass is zero.

The term 'scalar' simply indicates that, to describe the field at a point in space, a numerical value is sufficient (as opposed, for example, to a magnetic field which is 'vectorial' because it possesses a direction in addition to a magnitude, as shown by the alignment of iron filings in the vicinity of a bar magnet). The most obvious case of scalar magnitude is temperature, which does not have a direction, and a number suffices to define it. The wavicles of a scalar field are bosons, and this is what determines the special behaviour of the 37th field.

Like all the other fields, the Higgs field pervades the whole of space and has the same properties everywhere, whether in the centre of the Earth or in the dark and cold space that separates the galaxies. Differently from the other fields, however, that of the Higgs also *contains energy in the unperturbed quiescent state*, without external energy injected. As a consequence of this special property, combined with the fact that its wavicles (the Higgs particles) are bosons, the energy of the scalar field is *reduced* when space is filled with Higgs wavicles, rather than when it is empty. This special state is often called the 'Higgs condensate'.

Since every physical system tends to move to a state of minimum energy, empty space is therefore not truly empty, but filled with Higgs condensate; we can call it a 'filled-vacuum'. One can imagine it as made of extended Higgs wavicles, which attract each other and, being bosons, are 'happiest' when they overlap and form, in the whole universe, an all-pervading and uniform background medium. This medium does not change with time, because the interactions among the Higgs wavicles of the condensate make them stable; instead, a real Higgs wavicle produced in a collider is unstable and immediately decays in other particles.

Thus the 'vacuum' – which is the state of minimal energy without matter-particles and force-particles - is not empty but filled with such a Higgs condensate.[6] Since to pass from 'filled-vacuum' to 'empty-vacuum' (i.e. occupied by 37 fields *without Higgs condensate*) a large amount of energy is required, in the present day Universe the vacuum remains filled. But immediately after the Big Bang the temperature was very high and this was not the case: as we will see in the Epilogue, the transition from an empty-vacuum to a filled-vacuum happened *spontaneously* about 10^{-11} s after the Big Bang.

How can the existence of this 'condensate' produce the asymmetry between the masses of the photon and the neutral asthenon?

The reason is that photons and asthenons, moving in space, interact in different ways with the Higgs condensate. Photons are not affected at all by its presence, so the photonic wavicles simply continue to travel at the speed appropriate to massless physical bodies that transport energy, which is the speed of light. In contrast, the asthenons interact with the condensate – partially absorbing it – and these continuous interactions 'slow them down'; therefore they move at less than the speed of light, which is equivalent to having acquired a finite mass.

---

[6] Many physicists do not like the term 'Higgs condensate' applied to the 'filled' state of the vacuum, but I find this name useful to convey the idea that the Higgs field is in a different energetic state with respect to all other fields, which contain zero energy.

The difference is similar to the sensation one feels when using a fork and a spoon to stir honey in a jar; in the second case much more effort is required, as if the spoon had much larger mass. In this way, because of the interaction with the Higgs field, all the neutral asthenons acquire a mass of 91.2 GeV. The interaction of the scalar field with the charged asthenons is less intense, so the mass of the force-particles W^+ and W^- is 'only' 80.4 GeV. Photons do not interact at all and their mass is zero.

The Higgs Field and Particle Masses

Frank Wilczek, one of the founders of quantum chromodynamics, has explained how theorists came to the idea of the Higgs field.

> *Suppose that a species of fish evolved to the point that some of them became physicists, and began to ponder how things move. At first the fish-physicists would, by observation and measurement, derive very complicated laws. But eventually a fish-genius would imagine a different, ideal world ruled by much simpler laws of motion – the laws we humans call Newton's laws. The great new idea would be that motion looks complicated, in the everyday fish-world, because there's an all-pervasive medium – water! – that complicates how things move.*
>
> *Modern physics proposes something very similar for our world. We can use much nicer equations if we're ready to assume that the 'space' of our everyday perception is actually a medium whose influence complicates how matter is observed to move.*
>
> *This is a time-honored, successful strategy. For example: in its basic equations, Newtonian mechanics postulates complete symmetry among the three dimensions of space. Yet in everyday experience there's a big difference between motion in vertical, as opposed to horizontal, directions. The difference is ascribed to a medium: a pervasive gravitational field.* (Wilczek 2012)

The fish-physicist has difficulties in understanding the motions of things in water, which is nothing else than a *condensate* of water vapour, but would easily derive the laws of Newtonian mechanics by studying the movements in vapour, which is much thinner. Similarly the Higgs *condensate* complicates the behaviour of the other particles, endowing most of them with mass, while the Higgs field by itself would not hamper their motion.

At this point some metaphors can help to understand how the 'Higgs mechanism' gives mass to particles. The first is due to the Dutch theoretical physicist Martinus Veltman, Nobel laureate in 1999, who wrote in the magazine *Scientific American*:

> *Interacting with the Higgs field, particles acquire mass like strips of blotting paper absorbing ink. The pieces of paper represent individual particles and the ink represents energy, or mass. Pieces of blotting paper differing in size and thickness absorb different amounts of ink; in a similar way, different particles 'absorb' different amounts of mass. The observed mass depends on the capacity of the particle to absorb energy and the intensity of the Higgs field in space.* (Veltman 1986)

The second metaphor originates with the experimental physicist David Miller, and earned him a prize offered in 1993 by the British Minister for Science for whoever could successfully explain the Higgs mechanism in a way comprehensible to the general public, whose taxes were paying for LEP construction and running. The cartoons of Fig. 5.12 illustrate the essence of Miller's explanation.

Fig. 5.12 The metaphor used by Miller for the Higgs mechanism; at the arrival of Margaret Thatcher, many admirers crowd around her to ask for an autograph, so the movement of the past premier, 'weighed down' by the interactions with her fans, is slowed. If another, less famous, politician enters a smaller cluster is formed; he has a smaller mass and thus a larger speed in traversing the room

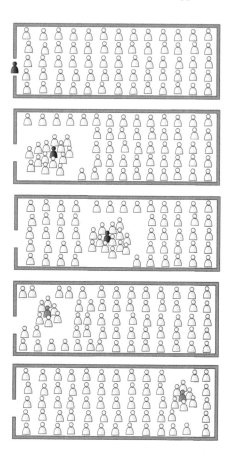

Imagine a cocktail party of political party workers who are uniformly distributed across the floor, all talking to their nearest neighbours. The ex-Prime Minister Margaret Thatcher enters and crosses the room. All of the workers in her neighbourhood are strongly attracted to her and cluster round her. As she moves she attracts the people she comes close to, while the ones she has left return to their even spacing. Because of the knot of people always clustered around her she acquires a greater mass than normal, that is, she has more momentum for the same speed of movement across the room. Once moving she is harder to stop, and once stopped she is harder to get moving again because the clustering process has to be restarted. In three dimensions, and with the complications of relativity, this is the Higgs mechanism. In order to give particles mass, a background field is invented which becomes locally distorted whenever a particle moves through it. The distortion - the clustering of the field around the particle - generates the particle's mass. (Miller 1993)

A third possible metaphor is that of a pair of twins, cross-country skiers, who cross a snow-covered 'field'. The photon-twin has waxed his skis well, while his asthenon-twin has not; therefore under the skis of the second twin a layer of snow forms which interacts with the piste and makes him move more slowly, as if his mass were larger than his brother.

If the Higgs field did not exist, therefore, a photon and an asthenon would behave in the same way; they would both be massless and have almost identical couplings.

For this reason one can speak of 'electro-weak unification'; the photon and the three asthenons are not mediators of different forces but only different facets of a single force-particle. This unification, however, is not easy to recognise by observing the effect of annihilations and of the collisions between matter-particles produced by accelerators, because the scalar Higgs field *breaks the symmetry*, making the electromagnetic force and the weak force appear as distinct entities, with mediators of very different masses and therefore with quite dissimilar ranges of action. Because of this the broken symmetry is often described as 'hidden'.

Up to now we have discussed bosons and seen that the omnipresent Higgs condensate gives mass to the asthenons but not to photons. With a different mechanism it also contributes to the masses of matter-particles, making the world in which we live varied and interesting. In fact, without the Higgs field electrons would not have mass and would also travel at the speed of light; in that case atoms and molecules, which make up all the matter in our universe, could not exist and form the physical basis of our consciousness. This is why physicists say that "*the scalar field fills space and breaks electroweak symmetry, giving mass to particles*" except photons, gluons and gravitons which do not interact with this field, and neutrinos, whose masses are minuscule and for which a different mechanisms has to be invoked.

Does the Higgs Field Really Exist?

Can it be demonstrated experimentally that space is occupied by a scalar field like the Higgs? To 'unmask' the field it is necessary to seek *evidence of its wavicles*.

Actually the Higgs condensate is the filled-vacuum through which everything travels; therefore it is not detectable with our instruments except through its unique effect, which is the mass of the fundamental particles. However, if the Higgs hypothesis is correct, injecting a sufficient quantity of energy will produce a localised oscillation, a wavicle which to measuring instruments will appear as a particle. This is the famous *Higgs boson*.

The Higgs boson has been at the centre of research in subatomic physics for many decades, not because it is important in itself but because its discovery would confirm the existence of the scalar field and the validity of the mechanism proposed to explain the mass of the asthenons and of the other matter-particles.[7]

---

[7] This is generally referred to as the 'Higgs mechanism'. However, it should be emphasized that citing only Peter Higgs too easily overlooks the other five theoretical physicists who all proposed

Thus, in the 1970s, each time a new accelerator began to operate, the availability of a higher energy made Higgs bosons of larger mass potentially observable, raising hopes of an imminent discovery. However, up until 2010, having studied millions of collision events in hundreds of different experiments, physicists had to admit that they had missed the target; either the Higgs boson did not exist or it had a mass larger than that which could be reached with the accelerators then available.

LEP and the Hunt for the Higgs Boson

The same thing happened when LEP, the Large Electron-Positron collider at CERN, started operating in 1989. Preceding results obtained at earlier accelerators said only that the mass of the Higgs particle must be larger than 30 GeV.

LEP began recording electron-positron collisions at about 100 GeV energy and it was foreseen to surpass 200 GeV in subsequent years. Since at LEP the Higgs boson should be produced together with a Z^0 asthenon (mass = 91.2 GeV) and at that time it was thought that the Higgs boson would have a mass below 100 GeV, we – LEP experimenters – were all convinced that sufficient energy would be available (200 GeV is larger than $91.2 + 100 = 191.2$ GeV). Therefore we were confident of finally observing, within a few years, at least a few tens of events which could unequivocally be attributed to the creation of the Higgs particle.

Each of the four detectors (ALEPH, DELPHI, L3 and OPAL) was operated by an international collaboration of about 500 physicists, originating in 30 or so laboratories from throughout the world. From the beginning of the 1980s the four teams were in constant collaboration and competition.

As we saw in Chap. 3, as a result of their low mass, electrons and positrons circulating in a synchrotron lose a fraction of their energy on each turn by emitting X-rays (synchrotron radiation). This loss of energy must be compensated by the kick from the oscillating electric field in the radiofrequency cavities. The more cavities there are, the larger the compensation for the losses and the greater the energy that the collider can attain.

Over the years the LEP energy was gradually increased, as one by one new accelerating cavities were installed, so as to reach 175 GeV. Not having found any Higgs signal, in the years 1995–1996 the French physicist Daniel Treille – my successor as DELPHI spokeperson – and other colleagues did everything possible to convince the CERN Directorate to invest about 100 million Swiss Francs in the construction of extra superconducting cavities, so that each of the two beams could reach 110 GeV. In these circumstances each electron-positron annihilation releases 220 GeV, an energy large enough to produce a Higgs boson of 130 MeV mass, which was predicted by the supersymmetric theory discussed in Chap. 6.

---

the same solution, in two independent publications, in the same year of 1964. These were Robert Brout and François Englert on one hand, and Gerald Guralnik, Carl Richard Hagen and Tom Kibble on the other.

But in those years the new LHC accelerator – which was to be assembled inside the LEP tunnel – was at an advanced stage of planning and required significant resources, both financial and in personnel; therefore the decision was finally taken to invest only in enough conducting cavities to reach 200 GeV.

At the beginning of 2000 we began to collect data at this energy knowing well that by autumn LEP had to stop to be dismantled and leave space for the LHC, which was to be built in the same 27 km tunnel. By October the energy had increased up to 209 GeV and the four collaborations were collecting data night and day and immediately analysing them in search of a signal.

In such circumstances the members of a scientific collaboration are bound to strict secrecy, but the CERN cafeteria is always crowded and rumours spread rapidly. So it became known that the ALEPH collaboration had observed one, then two, and finally three events, which could be attributed to the decay of a Higgs boson. L3 also observed a possible event, which raised hopes that the boson discovery was imminent.

At this point the CERN Director General Luciano Maiani was required to make a very difficult decision; if he were to delay by a year the dismantling of LEP, the thousands of people working on the new LHC project would lose enthusiasm and, possibly more important, CERN would have to pay penalty charges of more than 100 million Swiss Francs to external companies, already under contract for the dismantling of LEP and the installation of the LHC.

I well remember the long night-time telephone call that I made to Maiani to try to convince him to delay the closing of LEP. He listened to me patiently, as a friend, since we had known each other for a lifetime and considered ourselves almost brothers. But my call, as with all the pressure from hundreds of other friends and colleagues, did not budge him from his decision; so LEP was definitively switched off on 2 November 2000.

More than 10 years later Luciano Maiani wrote some beautiful lines about this difficult decision.

> It was necessary to kill LEP, the king of CERN, to build a larger giant, the LHC. I did it. There was much stress, which I feel as I write; it was really a transition drenched with great emotion. As well as a stubborn exercise of rationality.
>
> Lorenzo Foà and Gigi Rolandi from CERN urged me in a letter to prolong further the life of LEP. I could answer them with some justifications: 'The chance of finding ourselves by early autumn of next year still with only 3-3.5 sigma is not at all negligible.' I explained in the answer to my colleagues. 'In a limited time, no more can be done (or must we think of continuing for two years?). At this point we would have spent all our financial reserves, time and credibility on a very, very risky bet. I have never cared for poker.'
>
> For me – and not only for me – LEP had arrived at the end of the line. Only the LHC, with sufficient energy and luminosity at its disposal, could say if there was really a 114 GeV Higgs boson or if we were just chasing a ghost. (Maiani and Bassoli 2013)

A few weeks later, the ALEPH team (of which Lorenzo Foà and Luigi Rolandi were members) published a paper in the journal *Physics Letters* which concluded by claiming the observation of a "few tens of probable Higgs particles, with a mass equal to 115 GeV and a significance of 3 standard deviations". What is this 'standard deviation', indicated by the Greek letter *sigma*, and what did the 3–3.5 sigma quoted by Maiani mean?

The Rigour of Statistics

To be able to conclude that a certain number of Higgs bosons have been observed, it is necessary to analyse the detector signals from many particles produced by their decays. Track by track the properties of each particle are reconstructed – energy, mass, direction of motion – and it is essential to demonstrate that they are consistent with the products which should be found in the eventual decay of the boson being sought.

But the number of tracks left by particles and antiparticles produced in a collision event are tens, sometimes hundreds, and it may well happen that a certain configuration of tracks *only similar* to a typical Higgs decay is found; this is known as a 'background event' and in practice is indistinguishable from a 'real' event. For a given configuration of tracks, the 'real' events are always unavoidably mixed with background events; so analysis of the data must be carried out using statistical methods.

The 'standard deviation' is a quantity, which expresses the uncertainty (or 'error') which should be attributed to a measurement as a result of inevitable statistical fluctuations; in other words it is an estimate of the variability of the data due to random factors compared to the value which is being measured.

In our case, if for example 100 background events should be observed in a certain configuration of tracks, the standard deviation is simply the square root of 100, i.e. 10.

Suppose then that in an experiment 120 events are observed in that configuration; what does this mean? It could be that the excess of 20 events compared to expectation has been caused by the new phenomenon being sought but it could also be, on the contrary, that the background has fluctuated by *two* standard deviations and the hoped-for events are not present.

Table 5.2 shows that this eventuality is quite common because it happens – on average – *once* when the same experiment is repeated 45 times.

The "significance of 3 standard deviations" from ALEPH indicates therefore that the probability that the observed particles were not Higgs bosons was one part in 750 (third line of the table). From our daily experience, an error probability of 1 in

Table 5.2 The probability that the observation is due to a fluctuation diminishes substantially the greater the variation from the expected value, which can be expressed as the number of standard deviations

Number of standard deviations	Probability that the observation is due to a background fluctuation
1	1 in 6
2	1 in 45
3	1 in 750
4	1 in 32,000
5	1 in 3,500,000
6	1 in 1,000,000,000

750 appears really tiny. However, to claim a scientific discovery, physicists require that the data should have a significance of at least 5 standard deviations (and for good reason; the Higgs particle identified at the LHC actually has a mass of 125 GeV, not 115 GeV).

In the hunt for the elusive Higgs boson, LEP was therefore obliged to hand the baton first to the Tevatron at Fermilab (USA) and then to the Large Hadron Collider.

Chapter 6
The Discovery Announced, Susy and Beyond

Contents

© Springer International Publishing Switzerland 2015
U. Amaldi, *Particle Accelerators: From Big Bang Physics to Hadron Therapy*,
DOI 10.1007/978-3-319-08870-9_6

Denis Balibouse, AFP/Getty Images and Daniele Bergesio, TERA Foundation

The Main Auditorium *is on the first floor of the CERN main building and typical of many large lecture theatres. Constructed in 1954, until 2011 it still retained the original folding wooden desktops and the same large blackboards, on which erasures over the many years had left dim records of previous presentations.*

I entered it in awe in 1960 to follow lectures by three important personalities in the history of CERN: the Frenchman Jacques Prentki, the German Rolf Hagedorn and the Austro-American Victor Weisskopf. It was here, with some trepidation, that I gave my first important seminar, followed in later years by many others. Almost three hundred of us crowded into it during DELPHI collaboration meetings and it was filled to capacity when, in 1989, with the other spokesmen of the LEP collaborations, we presented results from the first months of operation.

The Main Auditorium has hosted unforgettable presentations by many of the great accelerator experts and physicists of the world. The one by John Adams, who in 1969 announced that the first proton beam had been accelerated by the CERN PS (Fig. 2.14 on p. 58), and another by Carlo Rubbia in 1983, in which he detailed the discovery of the W asthenon, remain particularly memorable. At the beginning formulae and equations were written on the three sliding blackboards, which later progressed to a mammoth slide projector; today it is sufficient to arrive with a USB memory stick to load a file onto the computer and project wonderful images and videos.

In December 2011 the two scientific leaders of the ATLAS and CMS detectors, the Italians Fabiola Gianotti and Guido Tonelli, showed the first indications of the existence of the Higgs boson, with a confidence level of 3 standard deviations. Then on 4 July 2012 Gianotti and the American Joseph Incandela presented data which confirmed the discovery. Among those present were four of the six theoretical physicists who in 1964 had proposed the scalar field as the solution to the problem of particle masses. In a Main Auditorium packed with researchers, the eighty-three year old Peter Higgs emotionally embraced Fabiola Gianotti and declared that he had never expected to see confirmation of this daring hypothesis during his lifetime.

Fermilab and the Golden Years of the Tevatron

The dismantling of the LEP magnets began in 2001 and the baton, in the search for the Higgs field, passed to Fermilab, the US Fermi National Accelerator Laboratory west of Chicago. It was here that the Tevatron, a proton-antiproton collider which was the first to use superconducting bending magnets, has operated since 1989.

The laboratory was founded in 1967 by Robert Rathbun Wilson, a physics professor at Cornell University where he had directed construction of various electron synchrotrons (Fig. 6.1). I shall speak more about 'Bob', as everybody knew him, in Chap. 8 since in 1946 he was the first to propose the use of proton beams for cancer radiation therapy. Here I limit myself to quote what was written in 2000, when he died, by his Cornell colleague and life-long friend Al Silverman.

Bob Wilson was a magician. Take, for example, his creation of Fermilab. This laboratory started as a proposal from the Lawrence Berkeley Laboratory for a 200 GeV accelerator to be built in seven years for $300 million. In due course the US Atomic Energy Commission (AEC) agreed to provide the funds but insisted that it should be built in Batavia, near Chicago. The Berkeley people refused, and the AEC turned to Wilson.

He had already gone on record criticizing the Berkeley proposal as too conservative, too expensive and too long. Wilson accepted the directorship and set out, more or less alone, to build the world's largest accelerator. So, with no staff, no laboratory and no buildings to work from, Wilson left for Batavia, a rural farm community in Illinois, to build the world's most sophisticated machine. One of Wilson's first acts in his new task was to reset the goals: instead of 200 GeV, the design was for 400 GeV; instead of seven years, the construction time was shortened to six; and the price dropped to $250 million. In six years *the machine was operating at 400 GeV and the cost was somewhat below the $250 million budgeted.*

How important was it to reach 400 GeV? After all, 200 GeV was already about a factor of six greater than any other accelerator. In fact, this target was crucial. The bottom quark could not have been discovered at 200 GeV.

That wasn't the end of the magic. He soon started a project, the Tevatron, to raise the energy to 1,000 GeV. But the tunnel's radius[1] was too small to use conventional iron magnets and the power for 1,000 GeV conventional magnets was prohibitive. The answer was superconducting magnets, the fields of which could be greatly increased and whose power requirements were reasonable. There was one small problem - no such magnet had ever been built. Undeterred, he set to work building them. He did this himself, with his sleeves rolled up and his hands dirty. Of course, he had talented people working with him, particularly Alvin Tollestrup. It was a prodigious accomplishment and it was the new higher energy that made possible the discovery in 1995 of the very heavy top quark. The third and heaviest family of quarks thus belongs to Fermilab and to Bob Wilson. (Silverman 2000)

In secondary beams produced by the Tevatron not only was the bottom quark discovered in 1977, but also the tau-neutrino was detected for the first time in summer 2000. Thus Silverman's statement can be enlarged: *"Three quarters of the third quark and lepton family belong to Fermilab and to Bob Wilson"*.[2]

In 1978, when he was 64, Bob Wilson stepped down as Fermilab Director to protest the lack of funding for the laboratory by the US Department of Energy (DoE). Leon Lederman – Nobel Prize winner as co-discoverer of the muon-neutrino – took on the job.

Following CERN successes in the transformation of the Super Proton Synchrotron (SPS) into a proton-antiproton collider, in 1982 Fermilab took the same route for the Tevatron. In 1985 physicists observed the first proton-antiproton collisions and, 10 years later, the collaborations which had built the Tevatron's two detectors, CDF (Collider Detector at Fermilab) and D0 (DZero), announced the observation of the production and decay of the top quark, predicted by the Standard Model but still missing.

During the following years Fermilab physicists and engineers greatly developed the techniques for 'cooling' the antiproton beam and succeeded in increasing the number of antiprotons accumulated in the Tevatron (Fig. 6.2).

With the increased luminosity the two detectors started to collect high-quality data in search of the Higgs boson. In 2010 a signal with a significance of 'only' 2.8 standard deviations was obtained, a result which the Department of Energy did not deem sufficient to justify further investment in the collider. Therefore the Tevatron was switched off in September 2011, a decision influenced by the commencement of LHC operation, which was already producing data at much higher energies and luminosities.

---

[1] This sentence refers to the tunnel built for the 400 GeV accelerator. Wilson called the new project the "Tevatron" because 1,000 GeV = 1 TeV (Tera-electron volt).

[2] The fourth member of the third family, the tau lepton, was discovered in 1975 by Martin Perl at the SLAC electron-positron collider SPEAR.

Fig. 6.1 Bob Wilson in 1994 at the reception for his eightieth birthday (Courtesy Fermilab Visual Media Services)

Fig. 6.2 The tunnel of the Tevatron is 6.8 km long. The Main Injector was built in the years 1993–1999 and has greatly contributed, with the upgrade of the antiproton source, to reaching the largest proton-antiproton luminosities ever obtained in a collider (Courtesy Fermilab Visual Media Services)

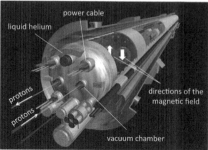

Fig. 6.3 The LHC tunnel and a cross-section through one of the 1,232 15-m long bending magnets. In the cross-section, the white arrows on the back indicate the opposite magnetic field direction in the two beam pipes, which are separated by 25 cm (Courtesy CERN)

Proton-Proton Collisions in the Large Hadron Collider

The LHC collider and its four detectors constitute the largest scientific and technical project ever carried out on Earth. Since 2010 the LHC has been colliding two proton beams. Because the particles of the two beams have the same electric charge, they cannot circulate in the same vacuum tube, as was the case for beams of protons and antiprotons either in the CERN SPS or in the Tevatron; instead they counter-rotate in *two separate* beam pipes of about 5 cm diameter, subject to oppositely directed magnetic fields (Fig. 6.3).

This greater complication with respect to a proton-antiproton collider is more than compensated by the fact that the counter-rotating proton bunches can contain many more particles, because protons are much more easily produced than anti-protons and no 'cooling' is needed. This implies that the number of collisions per second – i.e. luminosity – is about hundred times larger than in a proton-antiproton collider and so very rare phenomena, as the production of Higgs particles, can be recorded in much shorter running times.

In the LHC proton bunches are maintained for days on their circular trajectories by 1,232 very powerful bending magnets and constantly focused by 392 quadrupoles. The superconducting coils – cooled down to the temperature of about 2 K (-271 °C) by means of liquid helium – carry a current of 12,000 amperes (a normal copper cable would melt with a current ten times smaller). The two concentric LHC synchrotrons intersect at points where the four giant detectors for proton-proton collisions are located.

During 2010–2012 protons of 3,500–4,000 GeV were circulated, producing collision energies equal to 7,000–8,000 GeV; from 2015 it is foreseen to reach 13,000 GeV after a long shut-down during which the electrical connections between successive magnets will be made resistant to the larger currents circulating in the superconducting coils.[3]

---

[3] On 19 September 2008 – 9 days after the first proton beams successfully circulated in LHC – a poorly soldered electrical connection between two superconducting magnets led to the rupture of a

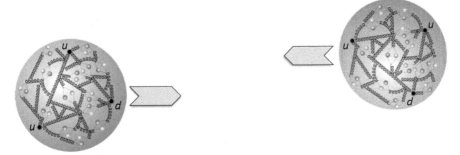

Fig. 6.4 When two high-energy protons collide, the *hundreds* of newly created particles and antiparticles are produced by the contemporaneous interactions of the three 'main' quarks, the gluons and the quark-antiquark pairs of one proton with the three 'main' quarks, gluons and quark-antiquark pairs of the oppositely moving proton

While physicists speak of collisions at 'high energies', in reality what they mean is 'high *energy density*'; each of the colliding protons actually has only the energy of a mosquito in flight but the density is enormous because in the LHC that amount of energy is carried by a single proton (while a mosquito is made of hundreds of billions of billions of protons and neutrons).

At this point I have to explain that in a proton-proton collision, as in a proton-antiproton collision, not all the energy imparted to the circulating particles is available for production of new particles and antiparticles.

The three quarks of a proton are bound by a continuous exchange of gluons, represented in Fig. 6.4 by the small springs. In turn these gluons transform into quark-antiquark pairs, which are created and then annihilated very rapidly, because their existence is limited by the Heisenberg uncertainty principle.

The complicated and ever changing internal composition of protons has been studied in detail in electron-proton collisions, produced in storage rings at the DESY laboratory in Hamburg (Fig. 3.14 on p. 83). The conclusions are astounding; in a proton which circulates in a high energy collider, *on average* half of its energy is carried by gluons. Of the remaining 50 %, 5 % is carried by quark-antiquark pairs and *only* 45 % by *u*- and *d*-quarks.

The energy carried by a *u*-quark or a *d*-quark is equal to 15 % of the proton energy, as obtained by dividing 45 % by 3; in other words a single quark has on average *one seventh* of the proton energy. Still less energy is carried by a gluon or an antiquark.

In the LHC collisions the effects of the distribution of the proton energy, among all its mobile and constantly varying constituents, are very important. For example, when in 2010 protons began to collide in the LHC with a total energy equal to 3,500 + 3,500 = 7,000 GeV, the available energy for a collision between a *u*- or a *d*-quark of

---

helium enclosure, causing damage to over 50 magnets and their mountings and contamination of the vacuum pipe, and delayed operations by more than 1 year.

Fig. 6.5 A proton-proton event registered by the detector of the CMS collaboration (Courtesy CERN)

the proton with a u- or a d-quark of the other proton was, *on average*, equal to only a seventh of 7,000 GeV, i.e. 1,000 GeV.

Fortunately this is only an *average* value because, sometimes, the two colliding quarks are randomly moving inside the two protons in such a way that the head-on collision between them liberates more energy than the average. This is the reason why it is often said that in the LHC the available energy in collisions between constituents is *ten* times larger than the 200 GeV of LEP.

In summary, protons are complicated objects, and colliding two of them to study their internal structure is – as Richard Feynman used to say – like hurling two Swiss watches against each other to shatter and understand how they are made inside.

The masses of the hadrons are well explained by quantum chromodynamics, as proven by the entries of Table 5.1 (p. 124). The same theory justifies the depiction of the proton by Fig. 6.4 and describes how, in the collisions of their point-like components, gluons, quarks and antiquarks – the many different components of the two protons interact simultaneously and *hundreds* of new particles and antiparticles are created (Fig. 6.5).

These processes are much more complicated than those occurring in the annihilation of dimensionless electrons and positrons, which produce a single flash of energy rapidly disappearing into the creation of only *tens* of particle-antiparticle pairs. Bruno Touschek (Fig. 3.2 on p. 67) with his inimitable humour once said: "*In proton collisions the uncouth snob shouts, and* one *understands little. In electron-*

positron annihilations, instead, we are dealing with particles which *speak civilly.*"
(Touschek 1977)[4]

The Detectors of the Large Hadron Collider

Six thousand or more physicists from the whole world have collaborated to build
the two enormous detectors optimised for the Higgs boson search, ATLAS and
CMS (Fig. 6.6), which are located 100 m underground. ATLAS occupies the
volume of a six-story building, which is a size about twice the linear dimensions
of the LEP experiments, because the energies of the hundreds of particles produced
in an LHC proton-proton collision are ten times larger.

ATLAS and CMS contain wire chambers which cover tens of thousands of
square metres and permit the reconstruction of the track of a charged particle with a
precision of a tenth of a millimetre. As at LEP, the wire chambers are organised in
concentric layers around the point where the collisions take place. The layers begin
a few metres from the vacuum pipe; closer to the collision point there are about ten
layers of silicon detectors which measure with micrometre precision the trajectories
of the charged particles which cross them.

The energies of the particles created at the collision point are deduced by measuring
the curvature of their trajectories, bent by enormous magnetic fields. To achieve this
both CMS and ATLAS have at their centre a superconducting solenoid, similar to
those of the LEP detectors but larger and more powerful. In addition ATLAS is
equipped with superconducting coils arranged radially (Fig. 6.7) to allow measurement of the energies of the muons which can penetrate many metres of iron.

In the LHC proton-proton collisions take place every 50 billionth of a second
leading to about a hundred million events each second; only a few hundred of these
can be registered by the computers for later analysis. To store just this tiny fraction
of the complex data produced, about ten million DVDs would be required each
year. Therefore it was necessary to build a new informatics infrastructure, the *LHC
Computing GRID*, a global computing network composed of thousands of computers, which allow storage of data and shared computing capacity to be distributed
throughout the entire world.

Events only just recorded at CERN can thus be analysed immediately by
physicists from all the universities and laboratories of the collaborations. Without
the GRID network it would not be possible to examine the enormous quantity of
data produced by the four LHC experiments; furthermore this unprecedented
capability for calculations and data processing will certainly have large technological repercussions in many other sectors. The development of the GRID involved

---

[4] As we know, the inconvenience of electron-positron annihilation is that these very light particles,
while circulating in a collider, lose energy by emitting light and – if one wants to have available
energies of the order of 1,000 GeV, as at LHC – they have to be accelerated not by synchrotrons
but by linear colliders (Fig. 3.10 on p. 78).

ATLAS

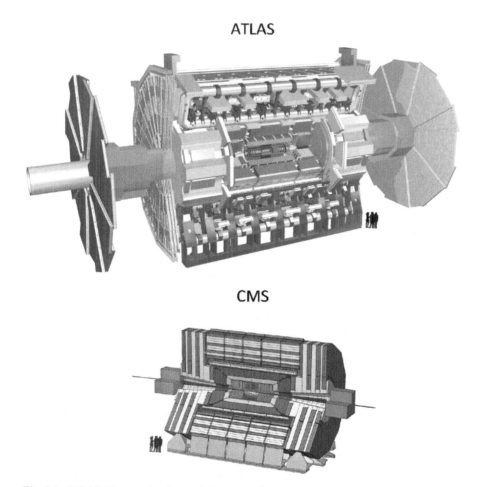

CMS

Fig. 6.6 ATLAS, illustrated at the top, is the largest detector ever constructed. The CMS detector, below, is smaller than ATLAS (the 'C' in the name stands for 'compact') but it still weighs more than 12,500 t

and will continue to involve companies working in information technologies; CERN is indeed a particularly interesting partner, because of its centralized organization with ramifications in all developed countries.

The GRID recalls another formidable CERN invention, which originated at the start of the 1990s: the World Wide Web. Created by the Englishman Tim Berners-Lee to allow physicists to rapidly exchange data over long distances, today it has become one of the drivers of economic development throughout the world.

Fig. 6.7 The eight superconducting coils of ATLAS, before the interior region was filled with tens of cylindrical detector layers. The figure in the lower part of the picture gives an idea of the scale (Courtesy CERN)

The Expected Discovery Is Announced

Forty years after the first prediction of its existence and after scores of lengthy experiments finding nothing, the elusive Higgs boson has finally fallen into the LHC net. This was not unexpected since, with the 1,000–2,000 GeV available in the collisions, it was computed that Higgs bosons having a mass as large as 800 GeV could be produced. This is a very comfortable ceiling, because in 2000 the four LEP collaborations, inserting into Standard Model calculations *all* the observations (and non-observations) which had been made, had concluded that the mass of the Higgs particle, if it did exist, ought to be in the range between 114 GeV and 200 GeV.

It required 2 years of data taking, but on 4 July 2012 the spokespersons of the ATLAS and CMS were able to announce the observation of a few hundred events which could be attributable to the decay of Higgs bosons, with a mass of about 125 GeV and a statistical significance of 5 standard deviations.

The Main Auditorium was packed, the atmosphere electric; tens of thousands of persons followed the presentations all over the world via the web, and there was much spontaneous applause during the presentations. The meeting was chaired by the German physicist Rolf Heuer, as CERN Director General. The media coverage exceeded all expectations, so much so that people spoke about 'higgsteria'. The phrase "CERN discovers the God particle" dominated headlines in newspapers and on web sites of every language.

This expression had become famous from the title of a book published in 1993 by the American Nobel laureate Leon Lederman (one of the three discoverers of the muon-neutrino), together with Dick Teresi. The choice of the title was more commercial than philosophical and mixed ideas which had nothing to do with each other; regrettably the name has become common currency.[5] If it really must be used, it would be more appropriate to apply it to the *field* rather than the *particle*; the scalar field could be called the 'God field' since, according to the Standard Model, it is the physical entity which pervades the entire universe and gives mass to particles and therefore has the function of "maintaining them in existence", something which theologians attribute to God.

Many interesting and original articles appeared in the world press during those days. I found one posted on the site of *The Guardian* newspaper by Tim Dowling, on the very day of the announcement, particularly amusing. It listed a set of instructions on how to explain the Higgs boson to different people, as in the following.

> For a child in the back seat of a car: "*It's a particle that some scientists have been looking for. Because they knew that without it the universe would be impossible. Because without it, the other particles in the universe wouldn't have mass. Because they would all continue to travel at the speed of light, just like photons do. Because I just said they would, and if you ask 'Why?' one more time we're not stopping at Burger King.*"
>
> For harassed, sleep-deprived parents: "*If the constituent parts of matter were sticky-faced toddlers, then the Higgs field would be like one of those ball pits they have in the children's play area at IKEA. Each coloured plastic ball represents a Higgs boson: collectively they provide the essential drag that stops your toddler/electron falling to the bottom of the universe, where all the snakes and hypodermic needles are.*" (Dowling 2012)

The Experimental Data

At the LHC the creation of a Higgs boson H^0 is usually due to the fusion between a gluon from one high energy proton and another gluon from the other colliding proton. Once produced, the boson H^0 decays in a large number of possible ways into lighter particles. The probabilities of each of the different decays, which can be calculated *exactly* by applying quantum field theory to the Standard Model, are listed in Table 6.1.

In general the decay probabilities diminish with a decrease of the masses of the matter-particles and antiparticles produced in the decay, because the lower the mass of a particle, the smaller the interaction of that particle with the Higgs field. Therefore the decay of a *b*-quark and a $\bar{b}$ antiquark (first row of the table) is much more probable than the decay into $\tau\bar{\tau}$ (row 2) which, in its turn, is more probable than the decay $c\bar{c}$ (row 3); but all these fermions decay in their own turn into many other particles and to identify them, among the scores of tracks which emerge from the collision point, is extremely difficult. The abundant decays into two gluons of row 4 of the table are also difficult to reconstruct. Thus, for the discovery, in 2011 and 2012 ATLAS and CMS studied with particular care the decay of the Higgs boson into two photons (row 5) and into four charged leptons (row 6) as shown in Fig. 6.8.

---

[5] The title was chosen by the publisher, who did not like the one initially given by Lederman to underline how much effort and money had gone and was going into its search: "The Goddamn Particle".

Table 6.1 The decay probabilities of a Higgs boson of mass 125 GeV. The last two decay modes in particular were the subject of the first LHC experiments

	Decay	In symbols	Probability (%)
1	In a $b\bar{b}$ pair	$H^0 \rightarrow b + \bar{b}$	65
2	In a $\tau\bar{\tau}$ pair	$H^0 \rightarrow \tau + \bar{\tau}$	5.5
3	In a $c\bar{c}$ pair	$H^0 \rightarrow c + \bar{c}$	2.0
4	In two gluons	$H^0 \rightarrow g + g$	5.7
5	In two photons	$H^0 \rightarrow \gamma + \gamma$	0.25
6	In four charged leptons	$H^0 \rightarrow e^- + e^+ + e^- + e^+$	0.008
		$H^0 \rightarrow e^- + e^+ + \mu^- + \mu^+$	
		$H^0 \rightarrow \mu^- + \mu^+ + \mu^- + \mu^+$	

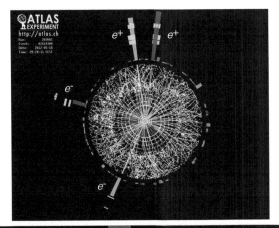

Fig. 6.8 Two events attributable to the decay of a Higgs boson in ATLAS (into 2 electrons and 2 positrons – i.e. into 4 leptons) and in CMS (into 2 photons). The light cones beside the symbols e^+, e^-, γ represent the energy of those particles (Courtesy CERN)

Fig. 6.9 Plotting the
number of events detected
at 7,000–8,000 GeV, with
4 leptons or 2 photons, as a
function of the mass M
(reconstructed from the
energies and directions of
the final state particles) both
ATLAS and CMS observe
peaks at the same mass
M = 125 GeV

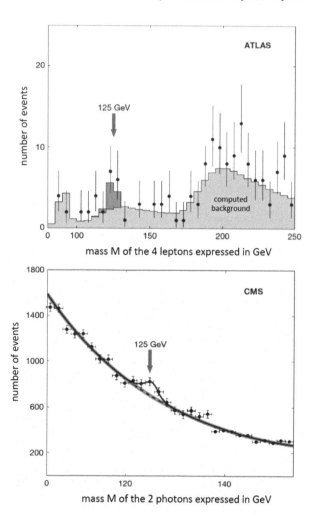

Fig. 6.9 Plotting the number of events detected at 7,000–8,000 GeV, with 4 leptons or 2 photons, as a function of the mass M (reconstructed from the energies and directions of the final state particles) both ATLAS and CMS observe peaks at the same mass M = 125 GeV

These decays have very low probabilities but are much easier to identify and permit a direct measurement of the mass of the particle which has produced them, because both detectors are able to measure with great precision the energies and directions of *all* the final particles.

The leptons and photons, decay products of a Higgs boson, are seen as tracks and localized energy depositions in the detector, but it should be noted that they appear in an event along with many tens, sometimes hundreds, of other tracks which tend to confuse the issue. As said in Chap. 5, in events in which no Higgs particle is created, the tracks due to other production phenomena can simulate the decay of a Higgs particle; these fake events are called 'background' events.

In Fig. 6.9 the numbers of events, which are consistent with the decay of a Higgs boson of a given mass into 4 leptons and into 2 photons, are plotted versus the mass. The points crossed by a bar are the *observed* numbers of events recorded in 2011 and 2012 by ATLAS and CMS, with their measurement errors, while the lines represent the *expected* numbers of background events. In both plots, around 125 GeV there are events in excess with respect to the background curves, indicating the production of a 125 GeV particle that decays immediately either in 4 leptons (top figure) or in 2 photons (bottom figure). Note that at the same value of the mass there are also many background events. For this reason, when an event which forms part of the peak is shown – such as in those illustrated in Fig. 6.8 – it is not possible to be *certain* if it should be attributed to a H^0 particle decay; one only knows that it *could be attributed* to the decay of a Higgs boson.

Following analysis of the signals obtained from several decay channels, both the ATLAS and the CMS collaborations observed – each independently of the other, and with a significance of 5 standard deviations – the production of a new type of boson with a mass equal to 125 GeV (it is certainly a boson because a fermion cannot decay into two photons, which are also bosons).

Does the new boson really possess the properties of the Higgs boson foreseen by the Standard Model?

In the July 2012 seminar the ATLAS and CMS representatives prudently stated that the observed decays were 'compatible' with those expected from a particle which has the characteristics of the Higgs boson. Indeed the measurements of the production probability and the decay probabilities reported by ATLAS and CMS were still imprecise.

In the following year, by collecting more events and refining the analyses, the measurement errors were reduced. The results of the two collaborations, properly combined, are shown in Fig. 6.10.

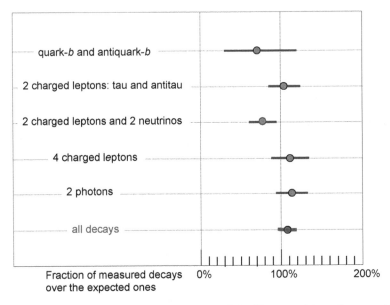

Fig. 6.10 The Standard Model predicts the decay of the Higgs particle to happen with the probabilities listed in Table 6.1 on p. 163. The figure shows– for five different decays – the ratio between the measured fraction and the expected value. If the theory is perfect and the number of events unlimited, all measurements would fall exactly on 100 % with very short horizontal bars

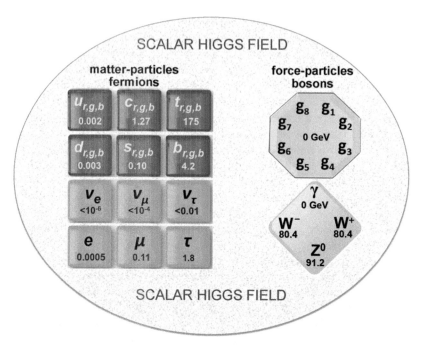

Fig. 6.11 This image illustrates the Standard Model after the discovery of the Higgs boson, and thus the scalar field that breaks the electroweak symmetry

Within the statistical errors – represented at the level of 1 sigma by the horizontal bars – which are still sizeable, the theoretical predictions are all fulfilled. Thus today one can confidently say that the new particle is the wavicle of the Higgs field which decays into the various channels as predicted and, in the Standard Model, gives mass to the other matter-particles and force-particles. However many years of LHC running and much more data are required to conclude, definitively, that all properties are *exactly* those predicted by the Standard Model.

If the new boson really is the wavicle of the field which – according to the Higgs mechanism – gives mass to the other particles, we can describe the Standard Model updated to 2013 with the diagram of Fig. 6.11. Here we have added to Fig. 5.1 of p. 125 a disc which surrounds the symbols of the 12 force-wavicles and of the 24 matter-wavicles, to indicate that they are immersed in a field which is present everywhere, with which they interact constantly, so acquiring mass.

The figure describes the Standard Model *qualitatively* and pictorially shows why physicists have been, and still are, so interested in the detection of collision events in which Higgs particles are produced. It is not to add a 37th particle to their already rich enough zoo, but because these particles are the direct proof of the existence of the scalar Higgs field, which in the figure is the gray platform, the ever present background, on which all the other 36 particles (and their antiparticles) move, interact and decay.

To perform *quantitative* calculations, the implied quantum field theory has to be supplemented with the numerical values of the masses of all the particles, of their couplings and of other quantities related to their decay probabilities. Overall, one needs about *twenty* 'parameters' that had to be measured experimentally because they cannot be deduced from the theory. Some of them are the masses of quarks and leptons, which – according to the Standard Model – are due to the interaction with the Higgs condensate.

Figure 6.11 shows that the masses of the lightest quarks are small: 0.002 GeV and 0.005 GeV for the quarks u and d respectively. This implies that about 99 % of the 1 GeV mass of a proton (or a neutron) is due to the kinetic energies of the quarks and to the energies of the gluon elastics that bind them, as depicted in Fig. 6.4, and *not* to the Higgs field.

Many physicists hope that – by studying the decays of Higgs particles – at some point discrepancies from the predictions of Table 6.1 will be found. That could then indicate the way towards a more encompassing theory since it would mean that the subatomic universe is more extensive and multi-faceted than that described by the Standard Model with *only a single* Higgs boson.

Going back to the LHC discovery, if further experiments will definitely confirm that it really concerns a scalar field which gives masses to the fundamental particles, I think the way will be open to a lexical change: the spread of the compact term 'higgson' – which sounds like 'fermion' and 'boson' – in place of the more clumsy 'Higgs particle'.

The last boson missing from the Standard Model has thus made its appearance in 2012, opening new avenues. This happened a little after the end of what scientists remember as the 'century of physics'.[6] In the year 1900, indeed, the German scientist Max Planck hypothesised that the energy of a light beam was transported by packets of indivisible energy, introducing the concept of *quanta*. It was the birth of the first boson, the wavicle of the electromagnetic field: the photon.

Supersymmetry and Superparticles

Even if the Higgs boson identified at the LHC has all the predicted properties, physicists will never claim that the Standard Model is completely satisfactory. Certainly it is not capable of explaining why the interactions between the Higgs field and matter-fields – which determine the enormous differences between the particle masses – are so different from one another.

---

[6] There are no *ifs* in history, *but* wondering *how* history might have changed if just one thing had been different is a favourite game of many science fiction writers and some historians. Joining the crew one can imagine the Higgs discovery announced in the year 2000 – exactly at the closing of the century – if only CERN had decided – when in the 1990s the technical possibility was there – to build enough superconducting radiofrequency cavities to push the maximum energy of the LEP electron positron annihilations to 220 GeV, thus increasing the range of explored masses up to 130 GeV.

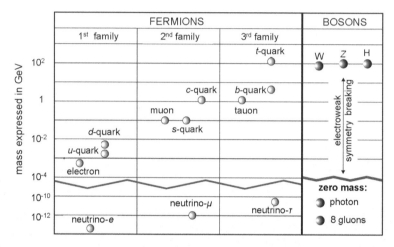

Fig. 6.12 The vertical axis represents particle masses on a logarithmic scale: passing from a horizontal line to the one above the mass is multiplied by 10. The *left* part of the figure shows that particles of the third family are heavier than those of the second, which themselves are heavier than particles of the first family. The neutrino masses are not yet well measured and the values represented in the figure are educated guesses

There are other reasons to believe that, with this discovery, particle physics is far from complete. Here I will describe just two of those reasons:

1. *The problem of mass*: if quantum field theory is applied to the Standard Model of Fig. 6.11, the values of the masses of the 36 force-particles and matter-particles can be explained but, at the same time, an absurd result is obtained: *the Higgs boson mass takes an enormously high value.*

2. *The problem of Grand Unification*: the data collected at LEP in the 1990s contain the message that *36 fields are not sufficient* to describe the subatomic world if, as well as electro-weak unification, unification of the strong force with the electro-weak force is required.

Both these problems would be resolved if new types of particle hypothesised by several theoretical models were to be observed at the LHC; these particles are so different from those of Fig. 6.11 to have earned the name 'superparticles'. Let us see first how the existence of these new objects would resolve the problem of mass.

The Higgs boson can transform itself – for a very brief interval determined by the usual uncertainty principle – into a virtual particle-antiparticle pair (electron-positron or muon-antimuon or quark-antiquark). Calculating all these processes with quantum field theory, it is found that the mass of the Higgs particle must be very much larger than the 125 GeV observed. Since the end of the 1970s the best physicists in the world have tried to find a more realistic prediction of the mass. The most satisfying solution has been found in the hypothesis that a *new form of symmetry between fermions and bosons should exist.*

In the description of data up to now, there appear to be two different worlds: the matter-particles (fermions) – at the left of Fig. 6.12 – and the force-particles (bosons) represented at the right.

Table 6.2 SUSY predicts
that for every fermion there is
a corresponding
supersymmetric boson, and
vice-versa

Fermions	Supersymmetric bosons
d-quark, *u*-quark	*d*-squark, *u*-squark
s-quark, *c*-quark	*s*-squark, *c*-squark
b-quark, *t*-quark	*b*-squark, *t*-squark
electron, muon, tauon	selectron, smuon, stauon
Bosons	**Supersymmetric fermions**
Higgsons	Higgsinos
Gluons	Gluinos
Photon	Photino
Asthenons	Astheninos
Graviton	Gravitino

In gauge field theories the fermions determine the existence of bosons, mediators of the forces, which is an evident absence of symmetry.

Supersymmetry – known familiarly by physicists as SUSY – hypothesises instead that every existing fermion corresponds to a new boson, and to every boson a new fermion. This hypothesis is at the same time both powerful and demanding, because it has as a consequence a physical world in which there is a complete symmetry between fermions (*exclusive* particles, whose wavicles each occupy their own individual state) and bosons (*inclusive* particles, whose wavicles superimpose with great facility, as happens to laser photons and the bosons of the Higgs condensate).

Since the Standard Model requires 24 matter-particles, 12 force-particles and a Higgs boson, the Supersymmetric Standard Model predicts another 36 superparticles and moreover – to be consistent – actually requires that there are *five* higgsons[7]! To distinguish these superparticles new names have been invented: for the boson superpartners of fermions the prefix 's' is added, while for the superpartners of bosons the names are modified to end in 'ino', as listed in Table 6.2.

We know that the masses of the particles observed so far do not exceed the 175 GeV of the *t*-quark. In the year 2000, when LEP was stopped, it was also known however that masses of the superparticles are certainly larger than 100 GeV, since at that time none of them had been discovered. It is therefore necessary in SUSY that a symmetry breaking mechanism analogous to the Higgs mechanism should exist, which makes each superparticle much more massive than the corresponding particle.

Thanks to the presence of new particles, in the supersymmetric model a Higgs boson has many more possibilities for transformation, since it can also create virtual

---

[7] In the so-called 'Minimal SUSY Model' the total number of fundamental fields is thus 77 (!) and their wavicles are 36 different types of 'normal' particles, 36 types of related 'super'-particles and 5 types of higgsons.

superparticle-superantiparticle pairs. If the contribution to the mass of the Higgs boson is calculated including these new virtual processes, it is found that this *exactly compensates* the enormous contribution due to the Standard Model particles – provided that the masses of the superparticles are not more than 1,000-2,000 GeV; with an arithmetic cancellation which is almost miraculous, a value for the H^0 mass compatible with the 125 GeV observed by ATLAS and CMS is obtained.

A further advantage of the SUSY model is that it predicts the existence of higgsons in a natural, actually an essential, way; electroweak symmetry breaking is not introduced *artificially* but originates in the theory without need of additional hypotheses.

There is even more. SUSY predicts not only that at least one Higgs boson should exist but also that the mass of the lightest higgson should be between 120 and 130 GeV! For this reason, many consider the 2012 discovery to be an indirect confirmation of the SUSY model. If however the mass were to exceed 130 GeV, the filled-vacuum of the universe would be in an unstable state and would decay at any moment into another type of vacuum, making every form of matter disappear. Here is why, after the discovery of the 125 GeV higgson, some journalists wrote that "the universe dances on the edge of an abyss".

By the end of 2013 – when LHC had to be stopped for repairs to the accelerator and to upgrade the detectors – no superparticle had been observed by CMS and ATLAS in the debris of the billion collision events recorded and very carefully scrutinized: the simplest minimal supersymmetric models do not apply and the mass of the lightest supersymmetric particles, if they exist, have to be larger than about 1,000 GeV.

This is troubling the nights of many theoreticians because the larger the mass of the superpartners the more difficult it is to explain the small Higgs mass with the miraculous cancellation quoted above.

Is it the time to despair? Many knowledgeable people do not. For instance Frank Wilczek – Nobel Prize co-winner in 2004 for the theory of quantum chromodynamics – has made a bet that superparticles will be detected by July 8, 2015, after new data will have been collected by ATLAS and CMS with LHC running – from the beginning of 2015 – at a total energy of $6,500 + 6,500 = 13,000$ GeV.

On this he wrote: "*I cannot believe the success (of the Higgs mass) is an accident. But in science faith is a means, not an end. Supersymmetry predicts new particles, with characteristic properties, that will come into view as the LHC operates at higher energy and intensity. The theory will soon undergo a trial by fire. It will yield gold – or go up in smoke.*" (Wilczek 2013)

If smoke is seen, we shall be reminded of Viki Weisskopf quoted at the beginning of Chap. 4 ("*The theoretical physicists are those fellows who stayed behind in Madrid and told Columbus that he was going to land in India*") and we would have to conclude that for too long SUSY has been the India of LHC.

The Couplings Depend on the Energy Exchanged

If Supersymmetry is realized in nature, a solution would have been found also to the second problem left open by the Standard Model, that of Grand Unification between the strong, electromagnetic and weak forces of Fig. 6.13.

To understand how this comes about it is necessary to recall that each force is characterised by the value of the mass of the force-particle and by its coupling, which is the probability that the force-particle should be emitted or absorbed by a matter-particle. The key point is that for each force *the coupling depends on the energy exchanged*. Figure 6.13 reports the values from Table 5.1, which however are valid only for $E = 0.5$ GeV; at other energies, the coupling changes. Let us see why.

Every virtual mediator of energy E has an influence for an extremely brief time within a certain small spatial volume, centred around the source-particle which emitted it, as we saw at the end of Chap. 2. The radius R of this volume depends on the energy of the virtual force-particle and is calculated from the uncertainty principle: $R = 0.2/E$, in which the energy of the virtual photon is measured in GeV and the distance in fermi. For example, a photon of 0.002 GeV can travel away only $0.2/0.002 = 100$ fermi, before being reabsorbed by the electron. Figure 6.14 illustrates how doubling the exchanged energy reduces by half the radius of the volume which is briefly occupied by the energy of the virtual mediator.

Despite the adjective 'virtual' the fact remains that, for a very short duration, an energy E is concentrated within a radius R. The presence of this energy produces observable effects, in particular it influences the value of the coupling, and therefore the probability that an interaction occurs.

Let us start by considering the emission of a virtual photon of energy E by an electron. The appearance of the photon implies that, inside a volume of radius

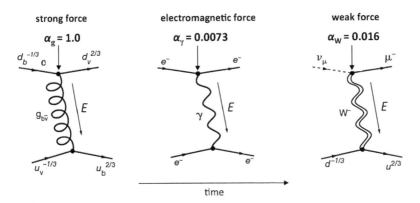

Fig. 6.13 The three fundamental forces and the values of the couplings, which are valid for an energy exchange $E = 0.5$ GeV

Fig. 6.14 Increasing the
exchanged energy reduces
the dimensions of the
volume in which the energy
is temporarily found. From
this it follows that the
coupling α_2 is different
from α_1

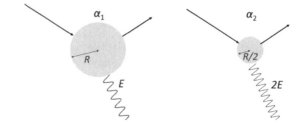

$R = 0.2/E$ around the source-electron, a concentration of energy temporarily appears. This energy sometimes creates a particle-antiparticle pair, which immediately annihilates. If the energy E is large so too can be the masses of the particle and antiparticle which are created. Conversely, at *low* energy – and therefore at *large* distances from the source-particle – only the lightest particles appear and immediately disappear.

Because each electron and positron has a mass of 0.0005 GeV, the creation of an electron-positron pair can take place if the photon energy E exceeds 0.001 GeV; the pair can be found at a distance from the matter-particle source not greater than $0.2/0.001 = 200$ fermi. To create a muon-antimuon pair 0.2 GeV is required, and the phenomenon occurs up to distances of about $0.2/0.2 = 1$ fermi.

In general, the virtual particles and antiparticles which arise and disappear continually around a matter-particle are closer to the source-particle as their mass increases; if the exchanged energy rises, particles and antiparticles of ever increasing mass contribute to this frenetic activity of creation and annihilation, appearing within ever diminishing radii. All these *virtual* processes have *real* consequences in modifying the local environment in which the source particle is found. The result of this *environmental* change is that the coupling no longer has exactly the same value but depends on the exchanged energy, because the virtual photon is coupled to the real matter-particle through the environment that surrounds it. The same thing happens to all the force-particles of Fig. 6.13.

We already know that the *strong* force becomes more powerful when the distance increases, because the gluons exchanged – which interact by means of the triple gluon vertex of Fig. 5.11b on p. 139 – form a kind of elastic which binds the quarks and the antiquarks. This implies that the coupling of a quark to a gluon, which has a value $\alpha_g = 1$ for $E = 0.5$ GeV, diminishes with the growth of the energy exchanged, that is when the distance decreases. And indeed at LEP for $E = 100$ GeV a value eight times smaller was measured: $\alpha_g = 0.125$.

The opposite happens to the electromagnetic force, whose coupling is $\alpha_\gamma = 0.0073$ when the exchanged energy equals 0.5 GeV (Fig. 6.13) and *grows*

Fig. 6.15 The force exerted
on a charge which passes
near to an electron is
screened less – and
therefore is larger – than the
force on a charge which
passes at a distance

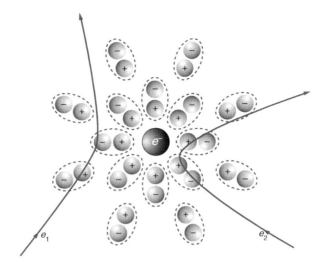

gradually with energy, becoming $\alpha_\gamma = 0.0078$ when the exchanged energy is
100 GeV (which corresponds to a distance of 0.002 fermi). This is the value
measured with great precision in the 1990s by the LEP detectors. Figure 6.15
explains how this happens.

An electron is always surrounded by pairs of virtual particles and antiparticles.
The figure shows that they are 'polarised', or oriented with the positive charge
closer to the negative charge of the source-electron. In consequence, a second
electron – a probe-electron – which passes close by feels a reduced repulsive
force, because the charge of the central electron attracts the positive charges and
repels the negative ones of the virtual pairs. Between the source-electron and the
trajectory of the probe-electron there are thus always more positive than negative
charges; they *reduce* the (negative) charge, i.e. the coupling, of the source-electron.

This 'screening' effect is less important when the probe-electron is passing
closer to the source-electron, because fewer pairs screen the negative charge of
the source-electron. In this case, corresponding to larger exchange energies, the
electromagnetic coupling is larger than when the distance is bigger and the
exchanged energies are smaller. Thus the ephemeral existence of virtual particle-
antiparticle pairs of larger and larger masses around an electric charge, when the
energy exchanged is increased, causes the increase of the electromagnetic coupling
α_γ from 0.0073 GeV at $E = 0.5$ GeV, to 0.0078 GeV at $E = 100$ GeV.

Grand Unification

Therefore with the growth of energy the strong coupling diminishes while the
coupling of the electromagnetic force increases; in other words the strong force
becomes *feebler*, the electromagnetic force *more potent*.

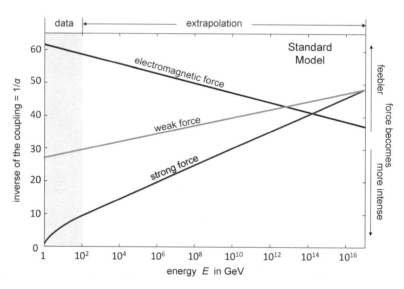

Fig. 6.16 The inverse of the couplings to matter-particles of the three fundamental forces as a function of the energy of the force-particle, according to the Standard Model

Does the *weak* coupling also 'run' and how?

The data collected at LEP in the 1990s confirmed, with much greater accuracy, what was already known from measurements made at previous accelerators; up to an energy of $E = 100$ GeV the weak coupling *diminishes* with the growth of energy, like the strong coupling. The experimental data are shown in the grey band at the left of Fig. 6.16. In this figure, the vertical axis shows the *inverse* of the coupling (because like this the resulting curves become almost straight lines); small numeric values therefore imply a large coupling, i.e. a more powerful force.[8]

The horizontal scale is logarithmic; at each tick mark the energy increases by a factor 100. The values of the couplings at energies greater than 100 GeV are calculated applying quantum field theory and using the measured values of the masses of all the particles and antiparticles of the Standard Model since, as we saw, the virtual particle-antiparticle pairs cause the variation of the couplings with the energy exchanged.

As can be seen from the figure, the couplings of the three forces tend to converge but 'miss the target'; around 10^{15} GeV they are almost equal, but the three lines *do not* pass through a single point. Therefore in the framework of the Standard Model, at no energy do the three forces have the same strength: 'Grand Unification' does not occur.

---

[8] For completeness it has to be said that in Fig. 6.16 and Fig. 6.17 the lines do not represent the electromagnetic and weak couplings α_γ and α_w, but two combinations of these quantities. This is a mathematical consequence of the unified electro-weak theory. For simplicity, in the plot the lines are indicated as 'electromagnetic force' and 'weak force' even if 'pure electromagnetic force' and 'pure weak force' would be better names.

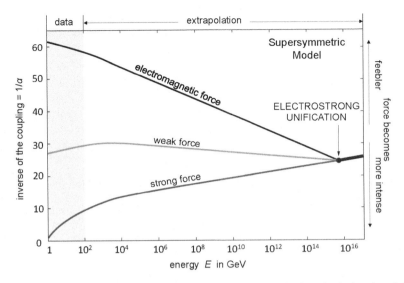

Fig. 6.17 The same figure as the previous one but this time using the hypothesis that the minimal supersymmetric model is valid and superparticles do exist, with masses between 300 and 3,000 GeV

An unification point can instead be found in the hypothesis that Supersymmetry is valid (therefore also taking account, in the calculations of the variation of the coupling, of the creation of virtual superparticles and superantiparticles). Figure 6.17 shows the couplings which we computed in 1991 with my friend and colleague Wim de Boer of Karlsruhe University and the PhD student Hermann Furstenau.

Using well-known equations derived from quantum field theory, we assigned masses to the supersymmetric particles so as to obtain convergence of the three forces at a single point. In this way we established that the masses of these superparticles should be between about 300 and 3,000 GeV. The new plot shows that in this energy range the three lines change direction – as a result of the virtual processes in which supersymmetric particles and antiparticles appear and disappear – and thus produce a unification point of the forces.

As we observed at the time, this convergence *does not demonstrate* the validity of the theory of Supersymmetry; however it is a clear indication of the *plausibility* of this elegant and revolutionary hypothesis, which is suggested by many other independent arguments and gives rise to *'electro-strong' unification* if the masses of the superparticles are not larger than about 2,000 GeV.

In subsequent years our approach has been improved, and often complicated, but the substance has not changed; without new particles and antiparticles with energies larger than several hundred GeV but smaller than 2,000–3,000 GeV (or without other new ingredients to add to the Standard Model), electro-strong unification cannot be obtained. Today SUSY is the most satisfactory theory capable of producing this important result.

Our work on the plausibility of Supersymmetry resonated strongly and in 2004 – when CERN had its 50th anniversary – it was one of the Top Ten articles most cited from among about ten thousand CERN publications. In his book *"Cosmic Imagery: Key Images in the History of Science"* the English astrophysicist John Barrow reproduced two plots, similar to Figs. 6.16 and 6.17, from our publication, and wrote:

> *In 1991, Ugo Amaldi, Wim de Boer and Hermann Furstenau showed that the cross-over just failed to happen unless a special 'supersymmetry', that had long been suspected to exist in Nature, existed. Its effect was to double the number of elementary-particle types in existence, and so slightly alter the way the force strengths change as energy increased. The result was a more-or-less exact cross-over at high-energy. This cross-over picture brought about huge interest in supersymmetric theories that continues unabated to the present day.*
>
> *The simple suggestive picture of the three-fold intersection of the couplings of the electromagnetic, weak, and strong forces of Nature – which was first drawn by Jogesh Pati in 1978 – has played an inspirational role in the search for an unified description of Nature. The convergence of the running force couplings suggests that unification does exist and led to the exploration of the early history of the Universe, an understanding of the preponderance of matter over antimatter within it today, and the search for the right way to include gravity in the unification scheme.[9] It is a simple symbol of the Universe's deep unity in the face of superficial diversity, which is what we mean by beauty.* (Barrow 2009)

This is another reason for which – as I said above – after the discovery of the Higgs field, the hunt for supersymmetric particles became the focal point of experimental studies at the LHC and the lack of positive results in the data collected in the first LHC years has disappointed so many physicists, including myself.

At present one can only say that until Supersymmetry will be confirmed by future LHC data with the discovery of superparticles of mass larger than 1,000 GeV, the convergence of the three curves in Fig. 6.17 will represent the most convincing indication of electro-strong Grand Unification, i.e. that, at exchanged energies of $10^{15} - 10^{16}$ GeV, the strong, electromagnetic and weak forces are different aspects of a single force, defined by *only one* coupling.

Strings

A supersymmetric theory of particles and the fundamental forces is aesthetically attractive, predicting the existence of the Higgs field, resolving in a natural way the problem of the masses of its wavicles and leading to electro-strong unification. However this is still only a partial unification because it omits the fourth force acting in nature: gravity.

A further unification will be extremely challenging, because it is very difficult to make quantum theory, which is at the foundations of the Standard Model, coexist with Einstein's General Theory of Relativity, which explains gravity as an effect of

---

[9] The early history of the universe, the matter-antimatter asymmetry and the unification of the gravitational force are discussed in the Epilogue.

Fig. 6.18 From crystals to strings, here represented as tiny loops which vibrate in three-dimensional space; string theory however is formulated in a space of ten dimensions and the strings can take many different shapes

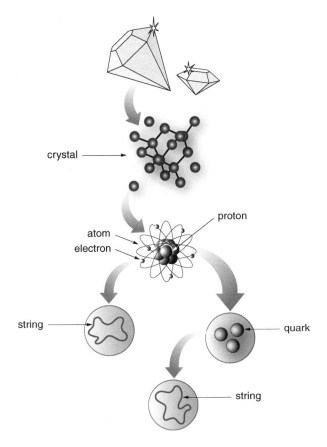

crystal

proton

atom

electron

string

quark

string

the curvature of space-time; to achieve it, it will be necessary to change some of their fundamental ingredients. In view of this, mathematical physicists have formulated many proposals but, so far, none of them is fully convincing.

I will mention only the theoretical construction which has enjoyed most attention in the last 30 years, still waiting for future experimental confirmation or refutation. This is 'quantum string theory' or simply 'string theory'.

While the Standard Model envisages the particles as point-like, this theory considers them as tiny and infinitely thin strings, or sometimes small loops, able to vibrate in many different ways, analogous to a guitar string emitting different notes depending on its length. Each mode of oscillation appears as a different particle; in one case an electron, in another a quark, as shown in Fig. 6.18.

Strings have lengths of the order of the *Planck length*, the smallest distance obtained by 'mixing' the fundamental parameters of quantum field theory and the universal theory of gravitation.[10] This dimension is only 1.6×10^{-35} m. In view of

---

[10] For quantum field theory the quantities are the speed of light c and Planck's constant, h. For the theory of gravitation, the only quantity is the gravitational constant, G. In 1899 Max Planck, the

the tiny value of the Planck length, it is not a surprise that strings appear point-like when compared to the dimensions of a proton, which is a hundred billion billion times larger.

In 1984 John Schwarz and Michael Green showed that string theory – which contains quantum gravity – can be made physically consistent with the Standard Model. This was the start of what was called the *first* superstring revolution.

But soon a large number of complications commenced.

First of all it was discovered that, to derive the Standard Model with the inclusion of supersymmetric particles, it was necessary that strings should not be structures in three dimensional space, which we know very well, but a space much richer in possibilities should exist, in that it has *ten dimensions,* one of which is the time dimension. According to the theory, the six dimensions of which we are unaware are 'curled up' on themselves at every point in ordinary space, so as not to be detectable even at the shortest distances which we can probe with the most powerful accelerators.

Secondly, it was quickly understood that hundreds of thousands of different ways existed to 'compactify' the six dimensions which should remain hidden. The possibilities proliferated and at the start of the 1990s five string theories had been identified, each of which allowed millions of ways to compactify the excess dimensions.

In 1995 Ed Witten launched the *second* string revolution by showing that the five theories are intimately linked by a new type of mathematical symmetry which he

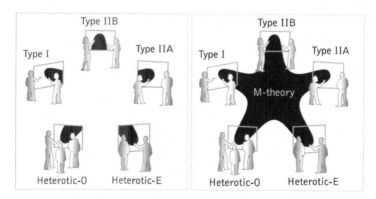

Fig. 6.19 In his year 2000 bestseller *"The elegant universe"* (Greene 1999), Brian Greene illustrated with these two pictures the discovery of 11-dimensional M-theory (**b**), which had previously been understood by theoretical physicists as five different string theories in a space of 10 dimensions, as shown in (**a**) (Greene 1999)

'inventor' of the photon, derived from these quantities the length that bears his name, the Planck length $= \sqrt{hG/c^3} = 1.6 \times 10^{-35}$ m.

called 'duality'. Moreover he conjectured the existence of a theory – called M-theory – in 11 dimensions, of which the five 10-dimensional string theories are special cases (Fig. 6.19). For this and other developments it was necessary to invent and construct a new type of mathematics.

Criticisms of String Theory

The M-theory is very general and still poorly defined; it is not possible to reconstruct the Standard Model, with its specific particles and forces, from it in an unambiguous way (even though the Standard Model can be made consistent with certain variants of string theory). However four of the five daughter-theories possess a very crucial property; each of its 'particles' must have its own 'superparticle'. In other words, these theories are *intrinsically* supersymmetric and the LHC discovery of at least one supersymmetric particle would be interpreted as a good indication of the validity of string theory.

Nevertheless it is improbable that this type of theory can be exposed to *direct* experimental verification, since the dimensions under consideration are of the order of the Planck length (1.6×10^{-35} m) that is much smaller than the spatial resolution obtainable even at the LHC (which reaches 'only' 10^{-19} m, as shown in Fig. 3.15 on p. 85). Today it is not possible even to imagine accelerators powerful enough to reach the Planck length, 16 orders of magnitude beyond present limits. Obviously it cannot be excluded that in some way – perhaps in many years time – *indirect* confirmation might be obtained.

Therefore the experimental prospects are slight and, moreover, the number of possible variants of string theory multiply year by year. This has given rise to some very critical opinions among many physicists, including Lee Smolin, who, in 2006, wrote a controversial book entitled "*The Trouble With Physics: The Rise of String Theory, The Fall of a Science, and What Comes Next*" (Smolin 2007).[11]

He is not alone in criticizing the enormous effort put into developing a theory, which is mathematically wonderful, but neither derives the Standard Model as the only viable theory of particle physics for energies smaller than 1,000 GeV, nor makes any direct contact with experimental data.

Already in 1987 Richard Feynman – the inventor of the exchange diagrams that bear his name which we have used frequently in this book – said: "*I think all this superstring stuff is crazy and it is in the wrong direction. I don't like that they don't*

---

[11] The number of string theory publications has grown enormously with time. In 1983 there were 16, around 50 in 1984, about 320 in 1985 and 640 in 1986. By the start of LHC data taking there had been about 50,000 papers by more than a thousand theorists. Their critics accuse the string theorists of having monopolised the field and prevented the development of alternative ideas.

check their ideas. I don't like that for anything that disagrees with experiment, they cook up an explanation – a fix-up to say 'Well, it still might be true.' When they write their equation, the equation should decide how many of these extra dimensions of space get wrapped up, not the desire to agree with experiment" (Feynman 1990).

The controversy has become even more acrimonious in view of the lack of evidence for superparticles at LHC and has contributed to revive several alternative theories. The strongest contender is 'loop quantum gravity' or LQG, which unifies quantum mechanics (with its uncertainty principle) with Einstein's General Relativity (with its space curved by the presence of mass and energy). The most important result is that space is *discrete*, made of grains of space, and quanta of the gravitational field, which have dimensions of the Planck length and are connected one to the other by ever changing 'links'. So LQG space itself, where all phenomena take place, is quantized and forms a kind of 'foam' where there is no need of superstrings (and of superparticles) to unify quantum mechanics with General Relativity.

After the end of the 2012 LHC run – during which ATLAS and CMS did not detect superparticles – Carlo Rovelli, who has made many contributions to loop quantum theory with Lee Smolin and others, has written:

> *It is as if the results from the LHC were speaking with the voice of Nature: 'Stop dreaming of new fields and exotic particles, extra dimensions, new symmetries, strings and all the rest. The data of the problem are simple: general relativity, quantum mechanics and the Standard Model. You 'only' need to combine them in the right way, and you will take the next step forward.' Nevertheless supersymmetric particles could exist at an energy scale still to be reached, and fundamentally could exist even if the loop theory is correct. Therefore, even if it is true that, while Supersymmetry has not exhibited itself where it was expected, and the string enthusiasts have gloomy faces and the loop advocates are smiling, one can still speak only of clues, not of proof.* (Rovelli 2014)

In spite of this, I am still convinced that strings, despite their difficulties and many uncertainties, remain the most likely candidate to justify broken supersymmetry, which in turn implies the unification of the electro-strong and gravitational forces, a possibility not even considered by quantum loop gravity.

Fermionic America and Bosonic Europe

Since 1948, when a beam from the Lawrence cyclotron bombarded a stack of nuclear emulsions and the first artificially produced pions were observed, particle accelerators have dramatically changed our vision of the subatomic world and the universe. Sixty-five years later, the Standard Model – based on solid principles of symmetry and on their hiding and breaking – has been confirmed to the smallest detail by the LHC. The discovery of the first higgson has opened the road to the unification of fundamental forces and convinced many authoritative physicists to believe in the theory of strings and in the broken symmetry between fermions and

Table 6.3 Proponents and discoverers of the particles of the Standard Model

Type	Particle	First proponent(s) or Laboratory where discovered	Year(s)
Fermions (fundamental matter-particles)	Electron	Cavendish Laboratory[a]	1899
	Electron-neutrino	W. Pauli, E. Fermi	1930, 1934
	Muon	Rome University[b]	1946
	Muon-neutrino	BNL (Long Island)	1962
	Tauon	SLAC (Stanford)	1975
	Tau-neutrino	FERMILAB (Batavia)	2000
	d-quark	M. Gell-Mann, G. Zweig	1964
	u-quark	M. Gell-Mann, G. Zweig	1964
	s-quark	M. Gell-Mann, G. Zweig	1964
	c-quark	BNL (LI), SLAC (Stanford)	1974
	b-quark	FERMILAB (Batavia)	1977
	t-quark	FERMILAB (Batavia)	2000
Bosons (fundamental force-particles)	Photon	M. Planck, A. Einstein	1900, 1905
	Gluon	DESY (Hamburg)	1979
	Asthenon W	CERN (Geneva)	1983
	Asthenon Z	CERN (Geneva)	1984
	Higsson H	CERN (Geneva)	2012

[a]The mass of the electron was measured by J.J. Thompson using a cathode ray tube and crossed electric and magnetic fields
[b]By using powerful magnets to deflect cosmic ray particles, Marcello Conversi, Ettore Pancini and Oreste Piccioni proved that the muon, the decay product of the pion, does not feel the strong force and is thus a heavier partner of the electron

bosons, which gives rise to the large *asymmetry* between the masses of the particles and the corresponding superparticles.

Looking back at the long road travelled one cannot fail to observe that the discovery of the particles which are the foundations of the Standard Model are subject to another unexpected form of *asymmetry*; since 1948 all the bosons have been discovered in Europe and all the fermions in America (Table 6.3). Up to today no theoretical physicist has proposed a convincing explanation of this surprising experimental fact.

Chapter 7
The Beginnings of Accelerators in Medicine

Contents

© Springer International Publishing Switzerland 2015
U. Amaldi, *Particle Accelerators: From Big Bang Physics to Hadron Therapy*,
DOI 10.1007/978-3-319-08870-9_7

Daniele Bergesio, TERA Foundation

First in her course at the Sorbonne when, in 1893 at twenty-six years of age, she was awarded her degree in physics. First, three years later, in the competitive examination for the extra qualification to teach physics. The first woman to be invited, in 1903, as a principal guest of the Royal Institution in London (though she was not permitted to be the speaker, because of her sex). The first woman to obtain a Nobel Prize, shared with her husband Pierre in 1903, "in recognition of the extraordinary services they have rendered by their joint researches on the radiation phenomena discovered by Professor Henri Becquerel". *The first female professor at the Sorbonne. The first scientist to be awarded a second Nobel Prize – this time in Chemistry – for* "her services to the advancement of chemistry by the discovery of the elements radium and polonium, by the isolation of radium and the study of the nature and compounds of this remarkable element". *The first person to be twice offered the Legion d'Honneur, which was both times refused. The first woman to be nominated, in 1922 when she was fifty-five years old, to the French Academy of Medicine, even though by a margin of one vote she was not elected to the Academy of Science.*

This is just a brief summary of some of the achievements of Marie Sklodowska Curie in the course of her scientific career, which at first sight seems to reflect a life studded with happy events. Her life was nevertheless badly upset by the death of her beloved Pierre, struck down by a wagon on 19 April 1906 when he was only forty-seven years old. Following this tragic event, Marie Curie began to keep a diary. Here we read: "On the Sunday morning which followed your death, Pierre, I returned to the laboratory for the first time. I tried to carry out a measurement, to complete a curve on which both of us had already obtained several data points. But I only felt the impossibility of continuing. On the way I walked as though hypnotised, without thinking of anything. I will not kill myself; I have no desire for suicide. But in the midst of all these carriages, why was there not even one which would take me to share the fate of my love?"

Great successes therefore, but also pain and great sadness, as her daughter Eva wrote in the final pages of her famous book "Madame Curie".

> The same picture, for me, always dominates the memory of these parties, of these occasions: the pale, almost indifferent, expressionless face of my mother. Willing or not, her personal prestige served to honour and enrich science – to 'dignify' it as the Americans would say – and she accepted that her legend was the agent of propaganda for a subject which was dear to her. But inside her nothing had changed; neither the fear of crowds, nor the self-consciousness which chilled her hands, which dried her throat, nor above all her incurable incapacity for vanity. (Curie 1938)

The Beginnings of Röntgen Rays in Medicine

With the discovery of X-rays in 1895, Wilhelm Röntgen originated a series of completely unforeseen developments in fundamental physics, but he also opened a new era in medical diagnosis and the treatment of cancer.

On 13 January 1896, 2 weeks after the announcement of his discovery, Röntgen demonstrated the extraordinary properties of his 'rays' – which today we know to be photons of energy around 10,000 electron-volts (0.01 MeV) – to Kaiser Wilhelm II. Only 3 weeks later, in Liverpool, the physician Robert Jones and the physicist Oliver Lodge used an X-ray radiograph to image a bullet embedded in a boy's hand. The exposure lasted four hours because the beam of electrons which generated the X-rays had a very weak intensity; indeed the 'Crookes tube' shown in Fig. 1.1 on

p. 4, despite being the best accelerator of its day, was able to produce only very few electrons.

Such a tube was a vacuum flask containing two electrodes with a high voltage between them. The few positive ions present in the residual air, attracted by the negative electrode (*cathode*), would strike it, extracting a small number of electrons; these, accelerated towards the positive electrode (*anode*), bombarded either the glass or the metal anode producing a very weak and unstable beam of Röntgen rays.

The radiographs were initially called 'cathodagraphs'; the name had an immediate success with the general public, such that in the 12 March 1896 edition of *Life Magazine* the following poem was published:

> *She is so tall, so slender and her bones,*
> *Those frail phosphates, those carbonates of lime,*
> *Are well produced by cathode rays sublime,*
> *By oscillations, amperes and by ohms.*
> *Her dorsal vertebrae are not concealed*
> *By epidermis but are well revealed.*
> *Around her ribs, those beauteous twenty-four*
> *Her flesh a halo makes, misty in line.*
> *Her noseless, eyeless face looks into mine*
> *And I but whisper: 'Sweetheart, je t'adore'.*
> *Her white and gleaming teeth at me do laugh,*
> *Oh! Lovely, cruel, sweet cathodagraph.*
> (Russell 1896)

Within a few months, X-rays began to be used in military field hospitals, to radiograph wounded soldiers. Already in October, Surgeon-Major John Battersby of the British Army assembled a small field hospital in Sudan, where hundreds of radiographs of soldiers wounded in battle during the Nile River Wars were taken. Another pioneer was the military doctor Colonel Giuseppe Alvaro, who in Naples radiographed two soldiers who had been repatriated after the battle of Adwa, fought by the Italian army against the Abyssinian army of Emperor Menelik.

In parallel to diagnostic applications, therapies were also developed. On 29 January 1896 Emil Grubbe, a student of homeopathic medicine at the University of Chicago, irradiated – but without success – the breast of a woman suffering from cancer.

In this period Grubbe was carrying out some experiments with the Crookes tube – as he recounted many years later in his book "X-ray treatment: its origins, birth and early history" (1949)- and on 27 January, having a very painful skin inflammation caused by X-rays on his right hand, he went to request a medical opinion from doctors at the Hahnemann Medical School in Chicago, where he was a student. One of those present observed that, according to one of the principles of homeopathic medicine, the new radiation, which was able to produce in large doses such damage to healthy cells, in small dose should be capable of killing cancerous cells. Another physician was so struck by this statement that he asked Grubbe to treat one of his patients suffering from an incurable breast tumour. Two days later Mrs Rose Lee was irradiated for an hour by a Crookes tube suspended 10 cm from the unhealthy breast, while the rest of the body was protected by layers of metal. As Grubbe wrote in 1933: *"without the blaring of trumpets or the beating of drums, X-ray therapy was born. Little did I realize that I was blazing a new trail, little did I realize that this was the beginning of a new epoch in the history of medicine"* (Grubbe 1933).

Despite having introduced the use of lead shielding himself, he absorbed very high radiation doses – like all radiation oncologists of that time – and lived to the age of 85 years while undergoing no less than 90 operations to remove radiation-induced tumours. Eventually he lost his left hand and forearm, most of his nose, upper jaw and lip so that, when people came to visit him, Grubbe would talk to them while keeping himself hidden behind a screen.

No scientific documentation exists of Mrs Lee's treatment and Grubbe's scientific priority has never been recognized by many, despite the acrimonious battles that he conducted all his life and the details given in "X-ray treatment: its origins, birth and early history". However one fact is certain: he became a very early ardent supporter of 'radiotherapy', which he taught with passion to physicians, and later in his life many honours were bestowed on him.

Instead, the irradiation of a patient sick with stomach cancer, carried out by the Lyon physician Victor Despeignes, is documented in an article of 26 July 1896. Since the patient died, many historians of medicine have however fixed the date of the origin of radiotherapy to be the following 24 November, with the first successful treatment carried out by the Viennese doctor Leopold Freund.

Freund was convinced that deep tumours could not be cured with the weakly penetrating X-rays of the period. Instead he used them to cure a skin tumour in a 5-year-old girl who had her back covered by hairy moles, and who, after the therapy, went on to live to the age of 70.

At the time Freund was working in dermatology; he chose to treat his young patient with X-rays because, as he wrote some years later, *"in June 1987 I read in a Vienna newspaper the joke news that an American engineer, who was intensively engaged in X-ray examinations, lost his hair because of business. This notice interested me very much"* (Mould 1993). He was the first to carry out radiotherapy with scientific methods; in 1903 he summarised his knowledge in a treatise, written in German, which was immediately translated into English with the title "Elements of general radiotherapy for practitioners" and has been continually updated and republished up to the present day.

Madame Curie, X-rays and the First World War

During the First World War, the use of X-rays saved many young lives. The experiences were dramatically successful in all countries and on all fronts, but I describe briefly only what took place in France, because it provides the opportunity to discuss an extraordinary figure, the Polish scientist Maria Sklodowska, better known as 'Madame Curie'.

Maria had moved from Warsaw to Paris in 1891 to complete her university education; here she met the French physicist Pierre Curie, then already well-known

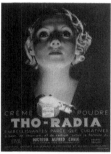

Fig. 7.1 At the beginning of the last century Marie and Pierre Curie's discovery had so caught the public imagination that radium was (dangerously) used as a component of many products, from black boot polish to beauty powder ((**a**) Courtesy David Pratt; (**b**) Musée Curie (coll. ACJC); (**c**) Alexandre Lescure/Musée Curie)

in international circles, whom she married in 1895. The story of the discovery of radium is so well known that I restrict myself to recall that it was the consequence of the decision taken in 1897 by Marie Curie, after the birth of her daughter Irène, to work towards a doctorate in physics research. After discussing possible subjects with her husband, she chose a new field – which she herself later named 'radioactivity' – where she could deploy her great talents as an experimenter.

In 1895 the French physicist Antoine Henri Becquerel had observed for the first time that uranium salts spontaneously emitted a particular type of natural radiation capable of registering images on a photographic plate. Marie chose to dedicate her studies to this new phenomenon, in collaboration with her husband; this arrived at the 1898 discovery of radium. Just six years later, in 1903, the Curie husband and wife team won the Nobel Prize together with Becquerel, and their names quickly became famous throughout the world (Fig. 7.1).

After Pierre's tragic death, Marie Curie strenuously continued her research work alone. Only the outbreak of the First World War induced her to interrupt her work, and it impelled her to dedicate herself completely to another valuable but demanding and dangerous activity.

Until then Madame Curie had never worked with an X-ray apparatus, but in 1914, soon after the German invasion of Belgium, she realised that there was an urgent need for radiological equipment on the battlefields, to help surgeons treat, and to extract bullets from the wounded. In a short time, using her influence with indomitable tenacity, she convinced several small companies to construct portable apparatus suitable for mounting on vehicles and personally undertook a long journey across France, Belgium and – towards the end of the war – also in Northern Italy, to educate doctors and nurses in the use of these instruments. Often she herself drove one of these special mobile X-ray units, of which by the end of the war twenty examples were in service, as well as more than 200 fixed stations (Fig. 7.2); more than a million wounded passed through these X-ray systems. It has been estimated that this initiative by Marie Curie, lasting from 1914 to 1918, saved the lives of thousands of French soldiers; and in the same way many more lives were saved among the lines of soldiers from other countries who were engaged in the trench warfare.

Fig. 7.2 (**a**) Madame Curie driving one of her mobile X-ray vehicles (World History Archive/ Alamy). (**b**) Radiological cars were invented at the beginning of the century but were considerably improved by Marie during the war and were soon called 'Petites-Curie', 'Little Curie' (Courtesy Conservatoire numérique des Arts et Métiers (Chalmarès 1905))

After the war Marie Curie wrote a book entitled "Radiology and War". Reading a few passages from it, one is struck by two aspects of her personality: the generous enthusiasm and the desire to remain in the shade, never speaking in the first person.

The story of radiology in war offers a striking example of the unsuspected amplitude that the application of purely scientific discoveries can take under certain conditions.

X-rays had had only a limited usefulness up to the time of the war. The great catastrophe which was let loose upon humanity, accumulating its victims in terrifying numbers, brought up by reaction the ardent desire to save everything that could be saved and to exploit every means of sparing and protecting human life.

At once there appeared an effort to make the X-ray yield its maximum of service. What had seemed difficult became easy and received an immediate solution. The material and the personnel were multiplied as if by enchantment. All those who did not understand gave in or accepted; those who did not know learned; those who had been indifferent became devoted. Thus the scientific discovery achieved the conquest of its natural field of action. (Curie 1921)

New X-ray Tubes

The speed with which the use of X-rays spread around the world, both in diagnosis and therapy, was due to two factors: newspapers and the radio daily informed the public, including physicians and physicists from the whole world, about scientific progress; in addition the Crookes tube was available in most physics laboratories.

In the first decade of the twentieth century the techniques to control, collimate and deliver the radiation dose were greatly improved. However, the duration of the exposure and the radiotherapy sessions remained very long with the Crookes tube. Finally, in 1913, the physician William David Coolidge developed and patented the tube that bears his name.

Will Coolidge was 18 years old when he enrolled at MIT – then called "Boston Tech" – in 1891 and he was 27 when he received his PhD at Leipzig; he then returned to MIT to work with the famous chemist Arthur A. Noyes. He was employed by General Electric in 1905 at its new research laboratory in Schenectady (New York), and within a few years saved the company from bankruptcy. Edison's patent on high resistance carbon filaments was threatened by inventions by other companies, which were encroaching on the booming market for incandescent light bulbs. Coolidge was entrusted with the task of studying tungsten, with the challenge of finding a way to make it ductile and therefore easily workable, so as to draw a new type of filament for incandescent lamps which would be more luminous, long-lasting and reliable. Coolidge applied himself to the problem for more than six years; he wrote, remembering that period, many years later:

> *Imagine a man wishing to open a door locked with a combination lock and bolted on the inside. Assume that he does not know a single number of the combination and has not a chance to open the door until he finds the whole combination, and not a chance to do so even then unless the bolt on the inside is open. Also bear in mind that he cannot tell whether a single number of the combination is right until he knows the combination completely. When we started to make tungsten ductile, our situation was like that.* (Suits 1982)

Eventually a process was developed by means of which tungsten – with trace metal additions – was made sufficiently ductile at room temperatures to permit drawing through diamond dies. As a result, General Electric began to produce light bulbs using filaments of tungsten, the material which has been used since then for all incandescent lamps. After this first important success, Coolidge continued working in the Schenectady laboratory to study the use of tungsten as a target for electrons accelerated in X-ray tubes (Fig. 7.3).

Fig. 7.3 This 1936 article shows photographs of the young William Coolidge holding one of his famous "tubes" and of him as an older man (*Springfield Sunday Union and Republican*, 21st November 1926)

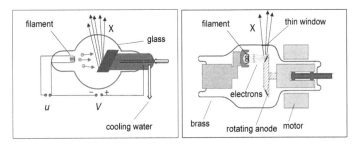

Fig. 7.4 (**a**) The internal structure of a Coolidge tube; on the *left* is the filament, on the *right* the anode that is cooled by water flow so that high currents can be achieved. (**b**) In modern tubes the heat is dissipated by rotating the anode

In 1913 the new tube was presented to physicians in New York, who immediately adopted it. There were two innovations; first of all the cathode, made of tungsten, was heated by an electric current and emitted a high and steady current of electrons, which were then accelerated towards the positive electrode; in addition the anode being struck by the electrons was cooled by a water flow, which prevented overheating even when the beam of X-rays was very intense.

In a tube the basic phenomenon is the emission of a photon by a fast electron that penetrates into the anode and, being deflected by an atomic nucleus, loses part of its energy in the form of a photon. The energies of the radiated photons range from zero up to the actual energy of the electrons; on average, the photon energy is only one third of the energy of the electron. For example, X-rays have a mean energy of about 0.03 MeV if the electrons are accelerated to 0.1 MeV. As shown in Fig. 7.4a, the photons directed towards the patient are the ones that emerge from the bombarded anode; the others are absorbed in the anode itself and contribute to its heating together with the energy lost by the slowed-down electrons.

With a Coolidge tube it was possible to produce more intense beams of X-rays and reduce the duration of the irradiation. It was only after 50 more years – in the 1960s – that the *rotating anode* was widely introduced which, by dissipating the heat more effectively, permitted the production of extremely intense X-ray beams. Effective heat dissipation is essential because 99 % of the energy of the electron beam goes into heating the anode and only 1 % is radiated in the form of X-rays.

Will Coolidge was not only a talented experimenter and a perceptive theorist but also a great manager. In 1932, during the Great Depression, he was appointed Director of the *General Electric Research Laboratory*, which he firmly steered both through the difficult period of staff reductions and the later developments and became the foundations of the GE of today. He also had unusual human qualities, as his biographer Chauncey G. Suits, who succeeded him as laboratory Director, writes:

Kindness and thoughtfulness in dealing with friends and associates were attributes that were deeply imbedded in his nature. I doubt if anyone ever heard him raise his voice in anger. His modesty was almost embarrassing, and he always viewed the accomplishments of his associates more generously than they themselves. He was greatly beloved by everyone who was privileged to be associated with him, and in the world of science, including medical science, he was regarded with deep reverence, as evidenced by the unprecedented award from the University of Zurich of a Doctorate of Medicine. Will Coolidge was blessed with remarkable health throughout his very active lifetime, and he retained a keen mind into his late nineties. He died on February 3, 1975, at the age of one-hundred-and-one. (Suits 1982)

Development of Tumour Treatments

The technical developments that made X-ray production easy and reliable came from the United States. During the same years the biological and medical foundations of the use of X-rays in tumour therapy were laid down in France.

In 1909 the Sorbonne University and the Pasteur Institute in Paris began the construction of a large laboratory for the study of radioactivity and its applications, realising a project strongly advocated by Marie Curie. The *Radium Institute* (today the *Curie Institute*) comprised two sections: the Curie wing, dedicated to research in physics and chemistry and directed by Madame Curie, and the Pasteur wing, for the study of the biological and medical effects of radiation and directed from 1913 by Claudius Regaud. The construction of these two buildings began in 1909 but was only completed at the end of the First World War. Madame Curie and Regaud together created the Curie Foundation in 1920 which, with an initial donation by Henri de Rothschild, constructed a dispensary in the vicinity of the *Radium Institute* which began to operate in 1922 (Fig. 7.5).

Fig. 7.5 In 1928 the President of the French Republic Gaston Doumergue visited the Radium Institute. From the *left* are Claudius Regaud, Marie Curie and Jean Perrin. Behind Perrin, Frédéric Joliot can be seen and, on his *left*, Irène Curie (Courtesy Musée Curie (coll. ACJC))

In Lyon in 1906 Regaud and Joseph Blanc had demonstrated, using rats, that the sensitivity to X-rays of the reproductive spermatogonial stem cells is at its maximum during the mitotic phase of cell division, while the most differentiated cells, in particular the spermatozoa, are less sensitive. This observation was transformed by Jean Bergonié and Louis Tribondeau into a law which carries their name: *"The effects of irradiation on cells are more intense the greater their reproductive activity, the longer their mitotic phase, and the less established are their morphology and functions"*.

Regaud also advanced the hypothesis that chromatin was the target within the cell sensitive to radiation; this was 50 years before the discovery of DNA, that is the fundamental constituent of chromatin in which radiation produces the lethal mutations. Also in Lyon in 1911 – aided by other skilful collaborators including Antoine Lacassagne – he had experimented on several animals and had begun to treat tumours applying the method of subdividing the radiation dose into multiple radiotherapy sessions (the technique of 'fractionation'); indeed, he had noted that fractionating the therapy into smaller doses rendered it more effective, because it gave healthy cells struck by the radiation the chance to self-repair.

During the First World War, Regaud had worked with the military diagnostic units instituted by Madame Curie and had himself organised several teams of surgeons to serve at the front-line. After the war, at the Pasteur Institute in Paris, he treated different types of tumour using the technique of fractionating the dose, mostly using radium sources which had been prepared in the Curie wing of the institute. The therapeutic method employed by Regaud is now called 'brachytherapy' but then was known, for obvious reasons, as 'Curie therapy'. Analogously, the treatment of tumours with X-rays was called 'Röntgen therapy' and is now known as 'radiotherapy' and less frequently 'teletherapy', since the source is far from the patient.

In brachytherapy a sealed radiation source, containing one or more radioisotopes, is placed at a small distance from the tumour to be irradiated. The radioactive substance can be placed directly in the tumour (interstitial method), in body cavities (endocavity method), or on the body surface (contact method). Today radium has been substituted by radioisotopes produced artificially in nuclear reactors or using cyclotrons, but the endocavity method is still employed for neoplasms resulting from cervical cancer of the uterus and also for other anatomical sites, such as the trachea, bronchi, oesophagus, bile duct, rectum and urethra. The interstitial method is widely used for prostate and breast cancer, while the contact method is reserved for skin tumours.

For the interstitial method, Regaud used platinum 'needles', which transmitted only photons emitted by the radium contained inside them, and invented an ingenious practical method to keep the needles in position during the treatment of cervical tumours. The treatments were administered in various Paris hospitals; Regaud and Antoine Lacassagne travelled by bicycle from one to another.

Claudius Regaud based his methods to treat patients on systematic irradiations of thousands of animals, from mice to rams. In 1927 he showed that it was possible to sterilise a ram's testicle without damaging the skin, provided the dose was subdivided into small daily fractions given over a couple of weeks. Regaud extended this result to the irradiation of tumours; he had certainly understood, from his own research in Lyon, that the testicle – with its continuous production of spermatozoa – is a good model of cancerous tissue with its uncontrolled multiplication of cells and that, in this case, the skin is the healthy tissue limiting the dose which can be administered to the target by radiation therapy. Today, those normal tissues which, close to malignant regions, cannot be irradiated without serious negative consequences for the patient's quality of life are called 'organs at risk' (OARs).

Starting from Regaud's pioneering studies, radiotherapists worldwide began to adopt dose fractionation, which was perfected in the 1930s by Henri Coutard treating tumours of the larynx with X-rays. Coutard began to work in the basement of the Pasteur wing directed by Regaud and then moved to the Röntgen wing of the Curie Foundation, where seven Coolidge tubes had been installed. With hindsight it is surprising that he arrived at the proposal of a treatment plan based on the fractionation into 30 sessions, still used today in radiotherapy, which at that time did not give good results because the electrons could not be accelerated to more than 0.2 MeV and at those energies the X-rays are insufficiently penetrating.

I would like to end this section by drawing attention to notable similarities between William Coolidge and Claudius Regaud, two scientists who, working 100 years ago on two different continents, made contributions to diagnosis and therapy which are still relevant and used today. They were both men of exceptional scientific, technical and managerial ability and gifted with similar human qualities, as one appreciates by comparing the passage cited at the end of the previous section with what has been written about Regaud:

> An exceptional versatile scientist, Regaud was comfortable in the laboratory as well as in the hospital wards and clinics. He also had great ability as an administrator. His awe-inspiring presence hid a charming, almost childish sensitivity. Obliged to exercise authority, he did it thoughtfully, with a great deal of consideration for the feelings of others. (Regato 1993)

Cobalt Bombs and Betatrons

In the first decades of the Twentieth Century, in parallel with the increasing understanding of the subatomic structure of matter in physics, biological research was uncovering the internal structure of the cells of living organisms. Thirty years after the pioneering medical studies of Regaud, it was possible to explain at a molecular level the effects of radiotherapy on cancerous cells. Nowadays we know that the effects are due to the energy deposited in the DNA by the electrons put in motion by the photons of the X-ray beam traversing a tissue.

These 'secondary' electrons travel in a zigzag manner in the tissue and ionise the atoms of the DNA molecules in the cells which they encounter, so suppressing the functioning of their genes, as we shall discuss in the next chapter. Since this is the mechanism, the energy deposited by the radiation is used to quantify the effect that a beam of radiation has on biological tissue. The unit of measure is the 'gray' (Gy) after the English physicist Harold Gray, who introduced it in 1940.[1]

Because DNA is contained in the cell nuclei, to give a *qualitative* definition of the gray it is useful to choose as reference a typical cell nucleus, with a mass of 10^{-10} g and 6 μm diameter; the *dose* of radiation absorbed by a sample of biological tissue is equal to a gray when 20,000 electrons are detached from the molecules contained in such a typical cell nucleus; the phenomenon is called 'ionization' because the molecule that has lost an electron becomes a positively charged 'ion'.

Until the 1950s X-rays for medical use were produced with Coolidge tubes, but the radiation was ineffective for treating deep tumours because, when a dose of 1 Gy is given to the skin of a patient, only 0.05 Gy is absorbed by tumour cells at 20 cm depth. To reach deep tumours, radiation oncologists increased the radiation dose by extending the duration of the exposure and irradiating with crossfire beams, to ionise more tumorous cell nuclei; however this reddened the skin and sometimes even caused skin burns.

In the 1940s tubes capable of accelerating electrons up to 1 MeV energy and producing more penetrating X-rays were constructed, but they did not become widely used. The first substantial advance in therapeutic techniques took place a few years after the Second World War, thanks to the introduction into clinical practice of a new apparatus called the 'cobalt bomb', a rather unfortunate name because it did not mean a weapon, but an instrument for healing. Essentially (Fig. 7.6a), it was a radioactive source embedded in a thick lead shielding and with an opening controlled by a shutter. The source was cobalt-60, a radioactive isotope with a nucleus made of 27 protons and 33 neutrons; while decaying with an half-life of 5 years the cobalt nuclei emit photons, each of which carries 1.2 MeV energy. When a patient is to be irradiated, an intense beam of these photons is allowed to emerge from the lead shield by opening the mechanical shutter.

These X-rays penetrate much further into tissue than Coolidge tube X-rays.

After cobalt bombs, the next step towards higher energy photons (and thus larger penetrations) was the introduction of circular accelerators called 'betatrons', which in the 1960s accelerated electrons for medical purposes up to 40 MeV (Fig. 7.6b). This brought another increase of dose with depth.

The story of betatrons is worth recounting. In 1928, Rolf Wideröe – who was an electrical engineer living in Oslo – wrote a PhD thesis on an invention of his, the first circular accelerator, the betatron, a kind of cyclotron that accelerates the electrons circulating in a magnetic field by induction, and not with a radiofrequency cavity as in a cyclotron. The betatron, which he called 'Strahlentransformator', was discussed – together with the description of a linac – in the paper we have already encountered in this book: it was the paper whose figures in 1929 gave Ernest

---

[1] Scientifically the *gray* (symbol: Gy) is defined as the absorption of one joule of radiation energy by 1 kg of matter.

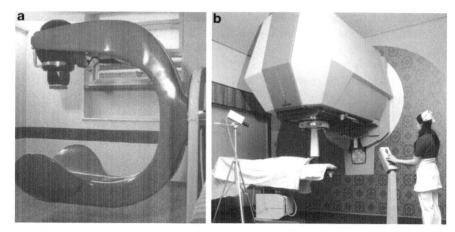

Fig. 7.6 (a) One of the first cobalt 'bombs' used in Italy in radiotherapy, in 1953 (Courtesy Ospedale Borgo Valsugana). (b) The 42 MeV betatron installed in 1972 at the University of Saskatchewan (Canada) was mounted on a gantry rotating around the patient (Courtesy University of Saskatchewan Archives, Harold Johns Collection)

Lawrence, who did not know German, the idea of the cyclotron – the second circular accelerator.

Only 10 years later a betatron accelerated its first electrons; it was built by Donald Kerst at the University of Illinois and commissioned in 1940. Stimulated by this achievement, Wideröe published a new paper on the design of a 100 MeV betatron, which was read by Bruno Touschek, who at the time was in Berlin. In 1943 Touschek wrote to Wideröe about some errors he thought he had found in the paper. He was invited to join the work on the betatron and moved to Hamburg in 1944, where he made important contributions to the theory of this new particle accelerator.

Bruno Touschek participated in the construction and commissioning of the betatron until his arrest by Nazi forces in March 1945, which I described in Chap. 3. Two facts from those months are relevant to the subject of this book. Firstly, as previously mentioned, Touschek and Wideröe debated the possibility of constructing particle colliders, a discussion that culminated 20 years later in Touschek's proposal of an electron-positron collider. Secondly, the two scientists realized the importance that high-energy photons – produced by the betatron beam when hitting a metal target – could have in tumour irradiation. In autumn 1944 this development was discussed in a meeting with Werner Heisenberg (Fig. 1.6 on p. 16).

Tumours and X-rays

To explain why electron beams produced by betatrons are definitely better than the ones due to Coolidge tubes, we have to make a detour and briefly discuss the origins of tumours and their medical treatments.

In a healthy organism cells multiply in a controlled way, which is only when the organism needs to grow or to replace dead or damaged cells. However, in our bodies there are always a certain amount of cells that are potentially cancerous; estimates vary between thousands and millions of precancerous cells in a healthy body. Their DNA has undergone many *mutations* and the genes, which regulate cell division, may act in an uncontrolled manner. Fortunately these cells are usually suppressed by our immune systems; however, when this does not happen they can multiply even when the organism does not require it. The tissue then grows out of control and is called a *tumour* or *cancer* (terms which we use synonymously).

More than a hundred types of cancer can affect humans. In many cases the tumour cells originating in one organ invade other tissues, dispersed to different parts of the body by the blood system or lymphatic system, giving rise to colonies called *metastases*. In other locations tumours grow without metastasizing but cause problems because of the increasing volume, particularly in the head.

Given the variety of affected organs and the pathological manifestations, to cure these illnesses (which annually strike about 4,000 people in a population of one million) a range of different and complementary methods are used.

The most direct method for 'solid' tumours, when they have well defined boundaries, consists in surgical removal of diseased tissue, which often also includes nearby lymph nodes, sites of possible metastasis. Radiotherapy is the second 'locoregional' methodology.[2] Each year about 50 % of all patients, i.e. about 2,000 people in a million, receive this type of treatment. It is often combined with *chemotherapy*, the administering of chemicals, which is a 'systemic' treatment targeted to destroy metastatic cells that cannot be removed by surgery or by radiotherapy.

Today radiotherapy, or treatment of tumours with X-rays, uses a penetrating beam of photons of 5–25 MeV energy to damage the genetic material of diseased cells and prevent them from multiplying. (It has to be stressed that radiation oncologists call 'X-rays' not only the photons produced in the collision of the electrons – accelerated up to 0.2 and 30 MeV in Coolidge tubes and betatrons respectively – but also the 1–2 MeV photons emitted by radioisotopes, which physicists call 'high-energy photons' and 'gamma-rays' respectively; in this book the oncologist style is mostly followed.)

The photons produced by electrons hitting a target have an *average* energy equal to one third of the electron energy. The photons of such an X-ray beam set in motion

---

[2] 'Locoregional' is a medical term meaning restricted to a localized region of the body where not only the tumour but also nearby potentially metastasised lymph nodes are irradiated. In the case of some solid tumours chemotherapy is also a locoregional treatment.

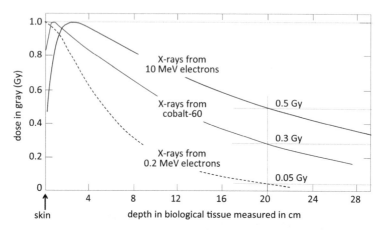

Fig. 7.7 The X-ray dose absorbed by uniform biological tissue is shown as a function of depth inside the material. The dose at 20 cm increases greatly when the X-rays originate from a cobalt-60 source, and even more when produced by 10 MeV electrons instead of the 0.2 MeV electrons in a Coolidge tube

some of the electrons of the traversed tissues, which lose energy ionizing some of the encountered molecules and come to a stop. X-rays are 'indirectly ionizing' particles because the ionizations are produced by the secondary electrons and not directly by the photons.

Figure 7.7 shows the depth dependence of the deposited dose in a biological tissue irradiated with the photons produced by a 0.2 MeV Coolidge tube, by a cobalt bomb and by a 10 MeV betatron. The average energies of the photons are in this case 0.07, 0.4 and 3.3 MeV respectively.

It is seen that by increasing the electron energy the dose deposited at 20 cm increases from 0.05 to 0.3 and 0.5 Gy. This fact fully justifies the efforts put along the years to increase the energy of therapy electron accelerators. In this process cobalt bombs and betatrons have been superseded by electron linacs, which have been built at the end of the Forties. They are discussed at the beginning of next chapter.

X-ray Diagnosis: From the Coolidge Tube to CT Scanners

Since the golden period 1895–1898 the applications of particle accelerators in therapeutics and diagnostics have been strongly interconnected – as depicted in Fig. 7.8 – with discoveries in fundamental physics.

Clearly the story of the betatron is a fitting example of how much the black and red yarns of fundamental physics and cancer therapy are intertwined.

Fig. 7.8 The discovery of X-rays, in 1895, and of radium, in 1898, gave rise to a continuous series of events in which discoveries and inventions in fundamental physics, made around particle accelerators, became entwined with repercussions in diagnostics and therapeutics

Untill now I have been following both medical yarns, but in the rest of this chapter I follow the blue diagnostic yarn. The next chapter is devoted to the red therapeutics yarn.

From its earliest use the Coolidge tube was applied not only in cancer therapy but also in diagnosis and today the X-ray tubes used for the now commonplace CT scan – or sometimes CAT, acronyms for 'Computerized Tomography' or 'Computerized Axial Tomography' – are still not very different from the original ones, which were low energy electrostatic accelerators capable of giving electrons kinetic energies of up to 0.2–0.3 MeV (200–300 keV).

In present-day diagnostic tubes, electrons of 50–100 keV (0.05–0.1 MeV) penetrate less than a thousandth of a millimetre into the tungsten anode, producing X-ray photons having an average energy one third of the electron energy. The photon beam is collimated to form a wide 'fan' of X-rays, 1 or 2 mm thick (Fig. 7.9).

During a CT scan photons directed at the patient interact with the electrons in tissue molecules and some are absorbed. According to which organs are exposed, a different flux of photons, which is then detected by sensors, emerges from the body of the patient. For example, the electron density in bone is one and a half times that of soft tissue; therefore the detectors reached by photons that have traversed the thorax and ribs measure a lower flux of photons than in the abdominal region.

The tube of the CT system moves slowly in a spiral trajectory, scanning the body of a patient lying on a couch. A powerful computer combines the values of the non-absorbed X-ray fractions – measured along all directions through the body – and reconstructs a three-dimensional image of the tissue; the bones, which are the most strongly absorbing, appear light while lung tissue appears dark. The result is a picture, which shows the shapes and sizes of the internal organs of the body (Fig. 7.10a), making pathological changes visible.

The development of computer assisted tomography earned a joint Nobel Prize in Physiology or Medicine in 1979 for Allan Cormack and Godfrey N. Hounsfield. The winners had done their work independently in the 1950s and 1960s in

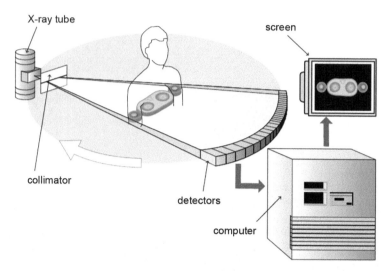

Fig. 7.9 The metal collimator of a CT scanner produces a thin and wide beam of X-rays which cross the patient's body. Photons that are not absorbed are measured by the detectors

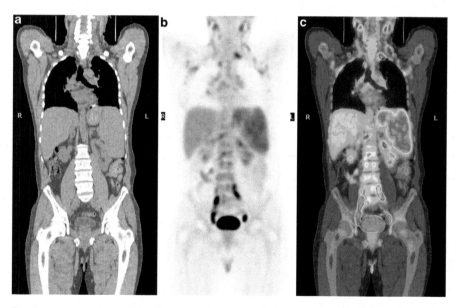

Fig. 7.10 (**a**) CT image. (**b**) PET scan of the same patient. (**c**) The merging of the two images highlights the distribution of the cancerous tissues in the left lung. (Centre Jean Perrin, ISM/Science Photo Library)

South Africa and in the United Kingdom respectively and did not know either each other or their respective researches (Fig. 7.11).

Cormack was a lecturer in physics at the University of Cape Town when, in 1955, the medical physicist at the Groote Schuur Hospital resigned. Since Cormack was the only nuclear physicist in the city, he was requested to spend three half-days each week at the hospital to supervise the use of radioisotopes. In this way he became interested in the problem of distinguishing tissues of different densities in the X-ray images. In subsequent years he worked intermittently on the mathematical problem of reconstructing the three-dimensional shape of internal organs of the body, by combining two-dimensional images obtained with X-rays originating from multiple directions, and he made several simple experiments. Despite the importance of the subject, the two articles written by Cormack in 1963 and 1964 for the *Journal of Applied Physics*, in which he described the fundamental ideas of computerised tomography, did not attract much interest. During his Nobel lecture Cormack commented: *"There was virtually no response. The most interesting request for a reprint came from a Swiss Centre for Avalanche Research. The method would work for deposits of snow on mountains if one could get either the detector or the source into the mountain under the snow!"* (Cormack 1979).

The English engineer Godfrey Hounsfield of EMI (Electrical and Musical Industries) ignored everything which had been previously done in this field when, some years later, he was struck by a sudden idea while walking in the countryside. During the Second World War he had worked on radar, the system which explores surrounding space by sending radio signals from a central point towards the periphery, by which the shape of objects within it can be reconstructed; during his walk it occurred to Hounsfield to ask himself whether it would not be possible to

Fig. 7.11 The 1979 Nobel Prize for Physiology or Medicine was awarded to Allan M. Cormack (*left*) and Godfrey N. Hounsfield ((a) and (b) American Institute of Physics Emilio Segrè Visual Archives, Physics Today Collection)

do the opposite, that is to study the contents of a box with beams of radiation which converge from the periphery towards the centre: "*I thought, wouldn't it be nice if I had many readings taken from all angles through a box and I could reconstruct in 3-D what was actually in the box from these random direction readings?*" (Wells 2005)

This apparently simple idea was difficult to realise both mathematically and in the necessary experimental verification. The experiments were made possible thanks to EMI financing. In 1967 Hounsfield finally succeeded in reconstructing the first three-dimensional images of inorganic objects and later the first CT pictures of animal brains. The first commercial apparatus – the 'EMI brain scanner' – was installed at Atkinson Morley Hospital in Wimbledon (London), where the first patient was examined in 1971.

An interesting anecdote is that in 1963, in the same period, EMI produced the first Beatles' record – 'Please Please Me' which was an immediate success, notably increasing the revenues of the company; it seems that this played a part, along with Paul McCartney's interest in Hounsfield's research, in providing financial support for the work and therefore its success.

Positron Radioactivity and PET Tomography

CT scans do not provide information on the metabolism of organs and their physiological processes, but only on their shape and dimensions; for example, they do not show irregular metabolic activity due to the development of cancer cells. For this it is necessary to have recourse to PET, or Positron Emission Tomography, which is an application of the discovery made in 1932 by Carl Anderson of the existence of a positive electron – a positron – among the stream of particles raining on earth from the sky (Fig. 1.9 on p. 19). This was the discovery, belonging to the 'fundamental physics' black yarn of Fig. 7.8, which 20 years later led to the first imaging instruments that made use of positron emitting radioisotopes. PET is one of the main spin-offs decorating the blue 'diagnostics' yarn.

To apply the PET technique a patient must be injected with a radiopharmaceutical tracer, which is a slightly radioactive substance able to enter the circulatory system of the organs to highlight the cellular activity within them.

The most widely used radioactive nucleus for PET scanning is fluorine-18, which is made by bombarding a water target enriched with oxygen-18, which contains more atoms of oxygen-18 than normal water, with a proton beam. Oxygen-18 has 8 protons and 10 neutrons and, when an accelerated proton strikes it, it absorbs the proton and loses a neutron, so becoming fluorine-18 composed of 9 protons and 9 neutrons. This radioisotope is made in an injectable form by chemically binding it to a sugar, deoxyglucose; the radiotracer is therefore called fluorodeoxyglucose or FDG.

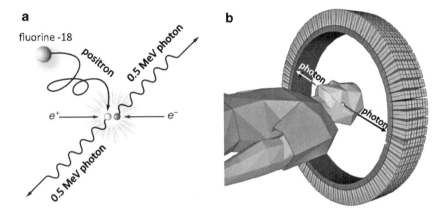

Fig. 7.12 (**a**) In PET the positron stops and annihilates with an atomic electron. (**b**) The two photons are detected simultaneously by two opposite elements of a scintillating crystal array

Tumour tissues have an intense metabolic activity, so they absorb more sugar from the blood than healthy tissues; in this way they also assimilate FDG and become charged with fluorine-18, which decays – with an average life of about 2 h – emitting positrons.

Each emitted positron travels a few millimetres through the surrounding tissue, slowing as it loses energy to the nearby atoms; finally it annihilates with one of the numerous atomic electrons present in the cells. The annihilation produces two photons, each having an energy equal to the mass of the electron and the positron, i.e. 0.5 MeV. These are the photons observed by the PET scanner; they travel in opposite directions (Fig. 7.12a) and are detected in a circular array of 'scintillating' crystals (Fig. 7.12b), made from a special transparent material, which emits a small flash of light when the photon gives up its energy.

A computer calculation reconstructs the line that connects the two opposite-facing crystals struck by the photons; this identifies the segment of the straight line on which the decaying fluorine-18 atom was located with a few millimetres precision. From the intersection of hundreds of thousands of such segments, the shape of the tissue volume in which the fluorodeoxyglucose was concentrated can be identified as a black area in Fig. 7.10b on p. 200.

The preparation and distribution of the 'labelled' sugar must be very rapid, because the number of radioactive atoms is reduced by half in less than two hours (the 'half-life' is 110 min). For this reason many large hospitals have their own cyclotron. In other cases the cyclotron is owned by a specialised company, which from early morning delivers vials to the hospitals. As long as the fluorine-18 radioactivity is not too much reduced, the product can be injected into the veins of patients during the morning.

FDG is produced by specially designed cyclotrons with energies in the range 8–19 MeV (called 'baby' cyclotrons), not much different from the ones built by Ernest Lawrence in the 1930s and 1940s, as we will see in the next section. FDG was originally developed in the 1970s by a collaboration of the US National Institute of Health, the University of Pennsylvania and Brookhaven National Laboratory, with the scientific objective of studying brain metabolism. Since then FDG radiochemistry has evolved fast and efficient manufacturing processes and today 75 % of all existing cyclotrons are devoted to FDG production. Applications include diagnoses of brain diseases like epilepsy and dementia, examination of heart functionality and detection of tumours.

Carbon-11, nitrogen-13 and oxygen-15 are isotopes which are much less used than fluorine-18, and also produced by baby cyclotrons. Other positron-emitting radioisotopes have very promising characteristics but need proton beams of higher energies to create them. Particularly interesting is gallium-68, which is made of 31 protons and 37 neutrons and has a 70 min half-life; it can be obtained from the decay of germanium-68 produced with a cyclotron of energy higher than 25 MeV. Chemically similar to iron, it can be easily attached to molecules that accumulate in neuroendocrine and other tumours. However this nuclear medicine examination is still not very common because of the dimensions and cost of cyclotrons that can accelerate protons up to at least 25 MeV.

If the two methods, PET and CT, are applied simultaneously and three-dimensional images are reconstructed, as in Fig. 7.10c, a PET-CT scan is obtained which provides information on both the morphology and the metabolic activity of the tissues, allowing very accurate diagnoses. The demand for PET-CT scanners, which combine metabolic and morphological imaging, is increasing by more than 5 % per year, as is the demand for positron-emitter radionuclides.

Cyclotrons for Radiotracing, Diagnosis and Therapy

To resume the threads of the story and describe the birth of the radiotracing technique known as PET, we have to return to the 1930s in Berkeley, to Ernest Lawrence's laboratory. Here the years that followed the invention of the cyclotron were very prolific in defining what would later become known as 'nuclear medicine'.

Following the arguments of Chap. 1, in 1932 Lawrence and collaborators were able to produce a 4.8 MeV proton beam with their 27-in. cyclotron. In 1934 two important discoveries influenced the future use of that accelerator: alpha-induced radioactivity by Frédéric Joliot and Irène Curie in Paris and neutron-induced radioactivity by Enrico Fermi and the other 'boys of Via Panisperna' in Rome. Those discoveries convinced Lawrence to dedicate its accelerated beams to the production of artificial isotopes.

Medical exploitation of the new radionuclides was undoubtedly in Lawrence's mind when in 1935 he called his brother John, a physician from Yale School of Medicine, to join him in Berkeley to study the beams and their artificial products. Although the mainstream activity of the laboratory was nuclear research, Ernest was personally very interested in medical applications and several cyclotron-

Fig. 7.13 Ernest Lawrence at the control of the 27-in. cyclotron together with his brother John (Courtesy Lawrence Berkeley National Laboratory)

produced radionuclides were used in studies of physiology and medicine, both in animals and humans. The foundations of the three applications of the radioisotopes still relevant today – radiotracing in biology, diagnosis in medicine, and endotherapy – were defined at that time (Fig. 7.13).

From the mid-1930s, radionuclides were introduced into animals and humans and traced in the internal organs with Geiger counters following physiological absorption and distribution. During these pioneering years of radio-tracing, the challenge was to find the right radioisotope that, as part of a molecule, would participate in the physiological processes and would be taken up, preferably, by a specific organ. The isotopes had to be chosen for their half-life and decay type, both to be compatible with the physiology of the processes under investigation and the detection procedure.

Radio-phosphorus-32 (which is made of 32 nucleons, i.e. neutrons and protons, and emits electrons with a half-life of 14.3 days) was discovered with the 27-in. cyclotron and used as a tracer to study the absorption and metabolism of phosphorus and cellular regeneration activity over several weeks.

In 1936 John Lawrence administered phosphorus-32 to a leukaemia patient for the first time, originating the therapeutic use of artificial radionuclides; this was the beginning of internal radiotherapy, which today is often called 'endotherapy'. Also in 1936 radioactive sodium-24 (half-life of 15 h), obtained at the Berkeley cyclotron, was one of the first artificial radioisotopes applied in physiology to study uptake and transport in animal and human circulatory systems and determine the speed of absorption. Radioactive sodium was first given to treat leukaemia patients in 1937.

Larger cyclotrons were constructed in the late 1930s at Berkeley, Harvard and Leningrad. Berkeley's 60-in. Crocker Medical Cyclotron was commissioned in 1939. As the name implies, this 220 t cyclotron (Fig. 1.12b on p. 25) was funded by the aged banker William Crocker, who was a Regent of the University and had in the past already donated personal funds for research. To convince him to finance the new apparatus, the President of the University of California Robert Sproul wrote to him: *"No other project the University could support has greater potentialities for the alleviation of human suffering. A Crocker Radiation Laboratory would be a lasting monument to the interest which your family has ever shown in public problems and the advancement of civilization"* (Heilbron and Seidel 1990).

Behind the moral pressure applied by Sproul, there were also more pragmatic reasons: Lawrence had threatened to leave Berkeley for the University of Texas where he had been offered an increased salary and a new laboratory. Sproul wrote to him: *"There is no more important business before the President of the University of California at this time than the resolution of any doubts you may have about continuing as a member of our faculty"* (Heilbron and Seidel 1990). In fact the funds arrived soon after.

For many years the pioneers of the new technologies were those in the Berkeley hills, but similar activities were already under way before the war in other laboratories; after the war, many cyclotron centres were established all over the world for biomedical research and other applications. Today many large hospitals have their own baby cyclotrons accelerating protons (and sometimes other light ions) to a maximum of 18–19 MeV. A survey of the production of the four major medical accelerator manufacturers estimated the number of cyclotrons operating in the world in 2010 to be 700, increasing at a rate of about 7 % per year in the years 2009–2012.

The technology of cyclotrons has greatly advanced since the times of Lawrence. The first area of progress is related to the shape of the iron poles producing the vertical magnetic field which bends the accelerated particles along a spiral trajectory (Fig. 1.11 on p. 24). The two oppositely-facing poles are no longer flat, as shown in Fig. 7.14a, but are made of 'hills' and 'valleys' so that there are regions (4 in the Fig. 7.14b) in which they are closer, and consequently magnetic field is larger. In such an Azimuthal Varying Field cyclotron the outward spiralling bunches of particles cross the boundary between a high-field region and a low-field region many times and are subject to alternating focusing and defocusing forces. As in all strong focusing accelerators, such an alternation causes an overall focusing effect on the bunches, which thus remain transversely compact.

The pole faces have the simple form of Fig. 7.14b when the maximum energy is about 10 MeV. At higher energies relativistic effects become important and the poles have more complicated shapes, as shown in Fig. 7.14c.

A second development is related to the maximum number of protons that can be accelerated every second (this quantity is called 'electric current' by physicists and

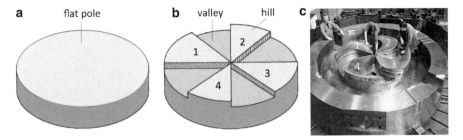

Fig. 7.14 (**a**) Two flat iron poles facing each other do not have any focusing effect on the circulating particles, (**b**) The sequence of hills and valleys produces a overall focusing effect. (**c**) The hills have a complicated shape in the cyclotron C230, which accelerates relativistic protons to 230 MeV for the therapy of deep seated tumours and is commercialized by Ion Beam Applications (IBA – Belgium) (Courtesy of IBA)

is measured in amperes). In the first cyclotrons the particle bunches were extracted with an electrostatic deflector (Fig. 1.11 on p. 24) and this process typically had an 80 % efficiency, so that 20 % of the current was lost and activated the deflector and the magnet. In the 1970s the expedient of accelerating negative ions was introduced. These ions are easily produced, because a *neutral* hydrogen atom is made of a proton and an electron; in a special gaseous source an electron can be added to a large fraction of the atoms, so that an outcoming beam of *negative* ions (H, made of one proton and two electrons) can be injected at the centre of the cyclotron and accelerated in the usual way.

The extraction of the beam is then simplified, since it can be achieved using a thin stripping foil conveniently positioned on the beam trajectory. In the foil the two electrons of each *negative* ion are stripped leaving the *positively* charged protons, which deviate in the opposite direction under the influence of the magnetic field and are extracted without the need of a deflector. The result is clean extraction with efficiency very close to 100 %, with a drastically reduced activation of the cyclotron components; in this way maintenance, regular inspections and decommissioning are considerably simplified. A remarkable characteristic of these cyclotrons is the possibility of obtaining simultaneous multiple beams by partial extraction using more stripping foils.

About 30 years ago several companies commenced production of baby hospital cyclotrons, machines that are stable, reliable and can run – in hospitals and radiopharmaceutical companies – continuously for days with accelerated currents of 50–100 microamperes, while requiring limited supervision and maintenance. The accelerator design has been adapted to requirements for radiopharmaceutical production, focusing mainly on radioisotopes for PET.

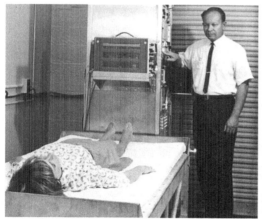

Fig. 7.15 Hal Anger (Berkeley) on the *left* and Powell Richards and Walter Tucker (BNL) on the *right* ((**a**) (Tapscott 1998) (**b**) Science Source, Getty Images)

Single Photon Tomography and Nuclear Reactors

Positron Emission Tomography is precise, but its instrumentation is more costly and less common in hospitals than detectors used in 'Single Photon Emission Computed Tomography' or, in short, SPECT. In SPECT a radioactive isotope is attached to a specific molecule and injected into the body, as for PET, but it emits a *single* high-energy photon (called 'gamma' by physicists), which allows location of the tissues in which the injected molecules are concentrated.

Other charged particles emitted by the injected radioactive nuclei, typically electrons, are stopped inside the patient's body, while most of the penetrating high-energy photons emerge without interacting and can be detected by one or more 'heads' made of scintillating crystals similar to the ones shown in Fig. 7.12b on p. 203. The detected photons come from only one direction because a block of lead with many small holes placed in front of the crystals stops all other photons. Hal Anger constructed the first instrument capable of making a two-dimensional image from the detected photons in 1952 at Berkeley and called it a 'gamma camera', which soon became known as the 'Anger camera'. By rotating two or more modern versions of such a device around the patient, the volume from which the photons are emitted can be reconstructed in three dimensions by a computer software (Fig. 7.15a).

The radioisotope technetium-99 m is used in 80 % of all nuclear medicine examinations and it is estimated that about 50 million patients in the world have a technetium examination every year.[3]

---

[3] In technetium-99 m the label 'm' indicates that this nucleus is 'metastable' and transforms into an almost stable nucleus with the emission of a photon.

The discovery of the element technetium is an interesting Italian story, which took place at the end of the 1930s in Sicily at the University of Palermo, where Emilio Segrè, coming from the Institute of Via Panisperna, had been appointed in 1935 as physics professor. He made a visit to Berkeley and Ernest Lawrence gave him an electrostatic deflector made of molybdenum, an element containing 42 protons, dismounted from the 37-in. Berkeley cyclotron. By analysing the deflector Segrè and the chemist Carlo Perrier found traces of an unknown radioactive element whose nucleus contained 43 protons. They later called it 'technetium' because it was the first artificially produced element, which had clearly been created by the particles that had bombarded the deflector in Berkeley over a long period. Shortly afterwards Segrè had to leave Italy because of Mussolini's racial laws and in Berkeley he discovered other technetium radioisotopes with Glenn Seaborg, including technetium-99 m. Segrè and Seaborg appear in the picture of Fig. 2.3 on p. 40.

Returning to medical applications of this artificial element, today in hospitals the chemical solutions that contain technetium-99 m and are injected into a patient's vein are not produced with a cyclotron but are extracted from a 'generator' containing radioactive molybdenum-99.

A generator is a system that contains a 'parent' radioisotope with a relatively long half-life, which decays into a radioactive 'daughter' isotope, characterised by a short half-life and immediately usable for the preparation of radiopharmaceuticals. The generator is schematically constituted by a cylindrical lead receptacle which contains a 'column' of special resin, into which is absorbed the parent radionuclide, manufactured by a nuclear reactor or cyclotron. From the radioactive decay of the parent, the daughter radioisotope is formed which, having different chemical characteristics from the parent, can be removed from the generator at the required moment by passing an alkaline solution through the column. This 'elution' process occurs without interfering with the bonding of the parent to the resin; in this way a phial containing only the daughter radioisotope is obtained. Since the parent atoms continue to decay, a large number of doses can be extracted daily from a single generator.

In the molybdenum/technetium generator the parent molybdenum-99, extracted from recently removed nuclear reactor fuel rods, is absorbed in an aluminium column. The molybdenum decays with a half-life of 66 h and the generator can be charged and despatched to all hospitals requesting it without significant reduction in its radioactivity. Then, within a week, the technetium-99 m, which has a half-life of only 6 h, is transferred to a suitable physiologically-compatible fluid passed through the column and, after a simple chemical treatment depending on the target organ, is immediately injected into a vein. The practicality of this method and the variety of molecules to which the technetium can be bound has determined the success of this form of Single Photon Tomography.

The first molybdenum/technetium generator was produced in the 1950s by researchers at Brookhaven National Laboratory, who then worked on the technetium chemistry in collaboration with scientists from George Washington University.

At that time, as well as the Cosmotron, BNL possessed a nuclear reactor employed for the separation of radioisotopes produced during nuclear fission and in the search for possible applications. Studying the separation of iodine-132, it was discovered that it was contaminated with technetium-99 m. Powell Richards, who directed the production of BNL radioisotopes, decided to employ this isotope for medical imaging (Fig. 7.15b): it had indeed a half-life which was neither too short nor too long, it could be bound to many molecules and the emitted photons, with an energy of 0.14 MeV, were easily detected by an Anger camera, or one of its more advanced successors. Walter Tucker and Margaret Greene brought the first generator into operation in 1958. Today technetium is employed for myocardial perfusion scans and for the scanning of bones, livers, lungs, kidneys and thyroid. I never understood why this important and useful series of discoveries and inventions was never recognized with a Nobel Prize.[4]

Will Cyclotrons Beat Nuclear Reactors?

In 2009 the two nuclear reactors at Chalk River (Canada) and Petten (Netherlands) – from whose uranium is extracted the major fraction of technetium used throughout the world – were out of action for several months because of serious technical problems, with the dramatic result that tens of thousands of patients could not undergo their indispensable SPECT tests. The crisis was very serious and since then studies of alternative production methods have been under way throughout the world. The use of a 50 MeV superconducting electron linear accelerator was proposed in Canada to produce uranium fission without the necessity of a nuclear reactor; for a machine of this type the necessary investment is large and the radiochemistry complex. In parallel, various companies planned the use of cyclotrons which accelerate high currents of protons of more than 20 MeV. For example, in 2013 the Canadian firm Advanced Cyclotron Systems requested permission for the use of technetium-99 m produced with its TR24 cyclotron, which accelerates a 500 microampere beam of 24 MeV protons; the technetium produced would be sufficient for a population of 5–7 million persons.

Looking today at the overall landscape of the use of radioisotopes in nuclear medicine, ideally medical cyclotrons should be evenly distributed throughout a region, with a large number of baby cyclotrons of energies smaller than 20 MeV, to fulfil requests for FDG and other PET radioisotopes; a sparse distribution of cyclotrons with wider range of energies – from 25 to 70 MeV – can produce all PET and SPECT radioisotopes used in less common diagnostic procedures, as well as the most commonly produced medical radioisotope, technetium-99 m.

---

[4] However, as we have seen in a previous chapter, Segrè won the Nobel Prize in 1959 for the discovery of the antiproton.

It is very difficult to predict in which direction research on medical radioisotopes will evolve in the long term. Many factors will determine the future of diagnosis with SPECT and PET, and the cure with endotherapy, which I have not discussed. Medical doctors, who today prefer to use one diagnostic modality, SPECT, and one radionuclide, technetium-99 m, will play an important role, with their future decisions, in the development of new accelerators and the substitution of this isotope with other radiopharmaceuticals.

A significant obstacle to the introduction of new techniques is the complexity of transforming an effective radioisotope into an approved pharmacological product. The most important factors to consider are the availability and cost of the raw material, the access to accelerators or reactors with appropriate energy and fluxes, and the existence of fast radiochemistry and the logistics to delivery the radioisotopes promptly before they decay. But one can be sure that cyclotrons will be used in the long term to produce radiopharmaceuticals used in diagnostic and also in tumour therapy, and it may even happen that the reactors, used for producing molybdenum/technetium generators, will be substituted by medium energy cyclotrons.

Chapter 8
Accelerators that Cure

Contents

© Springer International Publishing Switzerland 2015   213
U. Amaldi, *Particle Accelerators: From Big Bang Physics to Hadron Therapy*,
DOI 10.1007/978-3-319-08870-9_8

Chris Phan, Wiki Commons

Wilson Hall *is the central building at* Fermilab, *the particle physics laboratory near Chicago, the city where Enrico Fermi worked and taught from 1946 until 1954. This structure, more elegant than any of the CERN buildings, was desired by the founder and first director of the laboratory Robert Wilson, known by all – in the small world of particle physics – as 'Bob'. A visionary man, Bob Wilson was a sculptor, architect, physicist and accelerator expert, having studied at Berkeley as a doctoral student under Ernest Lawrence.*

He had previously directed the construction of several electron synchrotrons at Cornell University when, as described in Chap. 6, he was appointed in 1968 to create the new US national laboratory from nothing. In 1969, during a Senate hearing, he was asked about the value of the new accelerator to national defence.

His answer has remained famous: "It has only to do with the respect with which we regard one another, the dignity of men, our love of culture. It has to do with whether we are good painters, good sculptors, great poets. I mean all the things we really venerate in our country and are patriotic about. It has nothing to do directly with defending the country except to make it worth defending" (Wilson 1969).

Bob Wilson saw himself as a pioneer, a craftsman and a Renaissance man. These characteristics left their marks in the realization of the National Accelerator Laboratory (NAL), as it was initially called. One could see them in action during the construction of the laboratory – with its wonderful central building – and of the 400 GeV synchrotron.

When Wilson resigned as director in 1978 the CERN Courier *wrote: "During his term of office he has added the world's highest energy, highest intensity proton synchrotron to his previous similar achievement with the electron synchrotron in Cornell. Almost all features of the Laboratory (the aesthetic, the ecological, the hierarchical, the style of experiments. . .) are stamped with his powerful personality and he will not be an easy man to follow."*

Besides these important achievements, Bob Wilson was also the scientist who proposed – as early as 1946 – the use of proton accelerators in cancer therapy, and for this reason he opens this chapter.

Electron Linear Accelerators

As shown in Fig. 7.7 (p. 198), by augmenting the energy of the electrons the dose deposited by X-rays in a deep-seated tumour greatly increases with respect to the dose given to the surrounding normal tissues. To reduce the secondary effects of X-ray treatments, medical electron accelerators of higher energies have been developed throughout the years. In the last chapter we discussed the betatron of Wideröe and Touschek; in the 1950s these machines were superseded by safer, lighter and less power consuming electron linear accelerators.

The invention of the 'electron linac' arose from the fortuitous association of three friends; the American physicist William Webster Hansen who encountered two brothers of Irish origin, Russell and Sigurd Varian, during their university years. Bill Hansen was a physics graduate who in 1927, at only 28 years of age, became associate professor at Stanford University. Russell Varian also qualified as a physicist while Sigurd did not complete his college education and became a pilot, initially an enthusiastic amateur but later for several years a professional in central America. His experience of flying drove him to develop, with his brother's help, a special radio system for night-time position location of aeroplanes and of aerial obstructions.

The first attempt was a failure, so the brothers decided to seek help from their friend Bill, who was an expert in oscillating electromagnetic fields. In 1937 the three of them – working in a basement in the Stanford physics department, with $100 funding from the university – succeeded in constructing their first *klystron*.[1]

This can be described as an accelerator, with a first part which is similar to a Coolidge tube; the continuous flow of accelerated electrons in the beam is subdivided into bunches when traversing a special cavity in which a large electro-magnetic field oscillates. Such a cavity had been invented by Hansen who jokingly named it the 'rumbatron' (after the rumba dance, because the electromagnetic waves oscillate back and forth inside it). At the exit of the rumbatron, by passing through a second resonating cavity the bunches produced very brief, enormously powerful *micro*wave impulses; already at that time of about a megawatt, which is the electrical power consumed by a town with thousands of inhabitants.

The prefix 'micro' means that these waves are electromagnetic waves with a length of the order of 10 cm, short enough to be reflected by even small objects, and therefore well adapted to night-time location and identification of obstacles or other aircraft. Moreover the klystron of Hansen and the Varians was compact and light, so easily portable in an aeroplane, in contrast to the large devices for ships, which had been developed in parallel in the United Kingdom.

During the Second World War the three contributed to the development of military radar, but Bill Hansen's objective was to build an electron linear accelerator to carry out experiments on nuclear bombardment. He had been inspired by the idea of Widerøe illustrated in Fig. 1.10 on p. 23; an oscillating electric field, applied to a series of short tubes, gives an accelerating boost to a bunch of protons on each transit of a gap.

However electrons have a mass of only 0.5 MeV – about 2,000 times smaller than protons – and above about 1 MeV in energy they move, for practical purposes, at the speed of light. Compared to the case of protons, the oscillation frequency of the electric field must therefore be much higher; it is necessary to exceed a gigahertz, a billion oscillations per second. Hansen chose a frequency of 3 GHz (i.e. three billion oscillations per second) and produced microwaves both with a klystron and with a generator known as a 'magnetron' in which, applying the same principle of the klystron but adding a strong magnetic field, the electron beam followed a circular orbit. His first linac (Fig. 8.1a) was 90 cm long and accelerated electrons to 1.7 MeV in 1947; in 1950 with a 4.5 m linac electrons were accelerated to 6 MeV. Klystrons and linacs were readily exploited to produce X-rays for medical purposes. In 1948 the Varian brothers and Hansen founded Varian Associates, a company to manufac-ture klystrons and linear accelerators, which later became Varian Medical Systems.

In the same years in which Hansen and his collaborators were putting together the first linac, a second group of researchers was working strenuously on the same problem on the other side of the Atlantic, in the old Europe. In 1949, D.W. Fry and his team built a 40 cm linac, which accelerated electrons to 0.5 MeV, at the

---

[1] They did not know that in 1935 a paper describing a similar device had been published by an interesting couple of peripatetic physicists, the German Oskar Heil and the Russian Agnessa Arseneyeva.

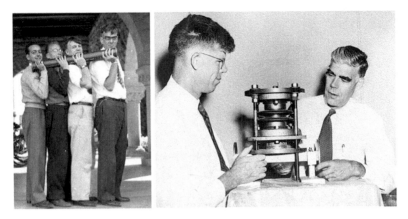

Fig. 8.1 (a) Bill Hansen with three students carrying a section of his first linac, which in about 3 m accelerated electrons to a maximum energy of 4.5 MeV (William Webster Hansen Papers (SC0126), Courtesy Stanford University Archives). (b) The two Varian brothers, Russell and (on the *right*) Sigurd, with one of their klystrons (American Institute of Physics, Science Photo Library)

Telecommunications Research Establishment in England. Not much later a 8 MeV linac was set up at Hammersmith Hospital in London, where the first patient was treated with X-rays in 1953; unfortunately the patient did not survive. Six months later at Stanford University the 6 MeV Varian linac irradiated the eye tumour of a child, who was definitively cured.

For the three principal American characters, their fine story typical of Silicon Valley ended tragically. Bill Hansen died at the age of only 40 years, of berylliosis, the terrible lung disease which he contracted from inhaling beryllium dust in experiments carried out during the war. In 1959 Russell Varian suffered a heart attack, at 61 years of age, during a trip to Alaska; 2 years later, at the same age, his brother crashed in Mexico piloting his own plane.

Today Varian Medical Systems is still the leading company producing radio-therapy equipment and, in particular, therapeutic electron linacs. In 2012 Varian Medical Systems had about a 50 % share of the total market, distributing yearly more than 1,000 linacs all over the world. This is a long-term consequence of the fact that radiotherapy equipment used today is based on a model patented by Varian Medical in the 1960s. The linac is mounted inside a support that rotates about a horizontal axis (Fig. 8.2b); 5–25 MeV electrons, accelerated by a linac, are deflected by a magnetic field and stopped by striking a metal target where they emit the photons which form the X-ray beam (Fig. 8.2a).

The radiation oncologist, with the aid of a medical physicist who produces a treatment plan calculated by computer, can choose the entry direction of the beam and its energy. By using multiple directions it is possible to deliver the necessary dose to the targeted tumour, and less to the surrounding regions, even when it is close to critical organs, which cannot absorb too much radiation without compromising the patient's quality of life.

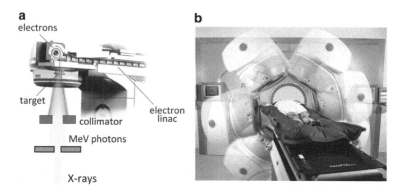

Fig. 8.2 (**a**) The transverse shape of an X-ray beam is collimated by a 'multileaf collimator', an assembly of metal blades controlled by a computer. (**b**) The apparatus rotates around the patient, projecting the radiation from different directions so as to minimise the effect on healthy tissues (Courtesy of ©Elekta, Stockholm)

As pioneered by Coutard and Regaud, a typical radiotherapy treatment plan foresees an average of 30 exposures of a couple of minutes each, to be carried out over a 6-week period. Using laser beams, a technician aligns the apparatus to the position of the patient, who lies on a platform, with a precision of a few millimetres. Then – during the few minutes of completely painless irradiation – the patient is left alone with only the hum of the machine and the irradiation headpiece, which turns automatically around them; the treatment room is occupied for 15–20 min.

Radiosensitive and Radioresistant Tumours

It is time to provide a more detailed, although still simple and often simplistic, explanation of the biological effects of X-rays.

X-ray photons are neutral particles which detach electrons from the nearby atoms of the tissue through which they pass, setting them in motion with energies equal to about half of their own energy. For example, a photon of 3 MeV – created by a 10 MeV electron in the target of an X-ray tube – gives up 1.5 MeV to an atomic electron, so that it detaches itself from the nucleus and acquires a speed equal to 94 % of the speed of light and, zigzagging around in the tissue, stops after 6–8 mm. The energy of this 'secondary' electron is used up interacting with the atoms which it encounters along its trajectory, either exciting them to a higher energy level or producing an ionisation, which means detaching another electron. In 6–8 mm of tortuous travel each electron produces about 40,000 ionisations.

These ionizations interfere with the normal functioning of many components of a struck cell and its surrounding water medium. Concentrating on the most important one, the DNA molecule, and simplifying many complicated phenomena, one often distinguishes *two effects* through which ionisations produced by electrons cause damage, sometimes lethal, to cells. By the *direct effect* the energy is deposited into a DNA double helix and, as a consequence, a weak chemical bond is broken in one of the two helices (a phenomenon known as a 'single strand break' or SSB), or, less commonly, on both strands (DSB, or 'double strand break'). The *indirect effect* refers to ionization of water molecules of the medium, which – following chemical transformations – become reactive 'free radicals'. These molecules wander throughout the cell for few millionths of a second and can eventually interact with the DNA molecule producing in it an SSB or, more rarely, a DSB.

In the nucleus of a cell that absorbs a dose of 1 Gy (by which the fast electrons cause 20,000 ionisations), around 500 single strand breaks are rapidly produced – either 'indirectly' by free radicals or 'directly' by charged particles that deposit energy in the double helix. DSBs are ten times less numerous, but they are the principal cause of cell death because, after some hours, SSBs are almost all repaired by molecular defence mechanisms which allow the cell to survive. Many DSB are also repaired, but this often does not happen to 'clustered' ones.

Accepting the approximate distinction between direct and indirect effects, one can say that *direct* effects are due to the *physics* of charged particle interactions with DNA, while those that are *indirect* are manifestations of the *chemistry* of free radicals. In the case of electrons set in motion by X-rays, roughly *80 %* of the lethal damage is due to *indirect effects* (therefore to chemistry) and only *20 %* to *direct effects* (i.e. to physics). The reason can be traced back to the fact that fast electrons are *sparsely ionizing*, because ionizations are about hundred nanometres (100 billionths of a metre) apart. This average separation is 50 times larger than the diameter of the DNA molecule so that a disrupting double or triple ionization on a short piece of DNA is a rare phenomenon.

Unrepaired or misrepaired DSBs do not cause the immediate death of the cancerous cell, which usually survives until the phase of cell division (mitosis). In this phase of the reproductive cycle, the radiation-induced DNA damage blocks the process and the cell perishes through a 'mitotic catastrophe'. In other words, the DNA damage does not kill the cell but either condemns it to death or blocks its division. Another mechanism, which plays a role in some types of cell, is *'apoptosis'*; in radiation-induced apoptosis irradiation activates a self-destruct mechanism in the DNA; this mechanism is important in healthy lymphocytes and in certain cancers, like lymphoma and leukaemia.[2]

A crucial condition required so that the indirect effects of X-rays can induce the death of a tumour cell is that it should be rich in oxygen. In fact, oxygen stabilises and potentiates the free radicals produced by radiation; if this element is insufficient, the free radicals rapidly disappear and behave less aggressively. In tumour tissue, whose vascular system is poorly developed, there are often 'hypoxic' cells (that are deprived of oxygen) in which the indirect effects due to free radicals are

---

[2] Other cell death pathways include 'senescence', where the cell stops growing and remains metabolically inactive, and 'necrosis', which leads to the disruption of the nuclear and plasma membranes.

reduced. Hypoxic cells are up to three times more resistant to X-rays than normally oxygenised cells; this factor is called the 'oxygen enhancement ratio' (OER). A large OER is a serious problem for radiotherapy with X-rays because it implies that, to eliminate the cancer cells, it is necessary to increase significantly the dose to tumour tissue, which unavoidably leads to more damage in the nearby healthy tissue. These tissues (called 'Organs at Risk', OARs) are actually what limit the dose that can be given to a tumour. A tumour can always be destroyed with a sufficiently high radiation dose, but this will lead to unacceptable damage to the surrounding normal tissues, which may be, in the short term, lethal for many healthy cells and can cause, in the long term, radiation-induced tumours.

Usually the normal cells are well-oxygenated, and therefore 'radiosensitive'; paradoxically, in the circumstances described above the destructive power of X-rays can do more harm to normal cells than to those which are cancerous, 'radioresistant' both because of the high OER as well as for other reasons intrinsic to the functioning of diseased cells. This explains one of the reasons for fractionating the dose, as a means to overcome the oxygen effects: subdividing the X-ray dose over 5–6 weeks, the progressive destruction of well-oxygenated cancer cells forces the hypoxic cells to move closer to the capillaries, making them radiosensitive, in a phenomenon called 'reoxygenation'.

A second reason for dose fractionation is linked to the fact that healthy tissue, whose cells multiply slowly, then has time to repair damage better than cancer cells. Another reason is that cells are less sensitive to X-rays when they are in a phase of the cell-cycle far from mitosis, and it is important that doses are delivered in such a way that all tumorous cells, which reproduce independently, are found – during at least several sessions of the treatment – in a phase of maximum sensitivity.

Using dose fractionation and combining radiotherapy with surgery and chemotherapy allows today to treat successfully most solid tumours. In this case, one speaks of 'radiosensitive tumours' but it is necessary to appreciate that the concept of radiosensitive tumours is not absolute, but relative to nearby healthy tissue, and that radiosensitivity depends both on the total dose delivered and on its fractionation.

There are tumours that are not radiosensitive, but no consensus exists on which fraction of tumours belong to this category. Pressed hard, many radiation oncologists would say that 5–10 % of tumours are 'radioresistant' mostly being hypoxic. They tend to resist irradiation with X-rays and represent an important medical problem to which many efforts have been and are devoted. To describe the approaches used today, we must return to Ernest Lawrence, his first cyclotrons and his brother John, the medical doctor.

Radiotherapy with Neutron Beams

After many trials on mice, in 1938 at Berkeley a human tumour was irradiated for the first time with a beam of 'heavy' particles, so-called because the neutrons used are 2,000 times more massive than the electrons accelerated in Coolidge tubes.

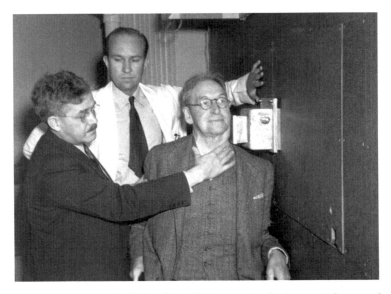

Fig. 8.3 John Lawrence watches Robert Stone aligning a patient in the neutron beam produced by the 60-in. cyclotron (Courtesy Lawrence Berkeley National Laboratory)

These fast neutrons were produced by colliding particles accelerated by the 37-in. cyclotron with a target.

The irradiation of animals was initiated in 1936 by John Lawrence and collaborators who compared results obtained with X-ray and neutron beams; these were the earliest contributions to the knowledge of the different biological effects of different radiations and their application to internal and external radiotherapy and to radiation protection. At the end of September 1938 the first patients were treated with fast neutrons. This first study on twenty-four patients, which used single fractions, was at that time considered a success and led to the construction of the dedicated 60-in. Crocker Medical Cyclotron, where neutrons were produced by 16 MeV deuterons striking a beryllium target.[3] Here the cancer therapist Robert Stone and his collaborators treated patients with fractionated doses using neutrons until 1943, when the cyclotron was appropriated for the atomic bomb programme (Fig. 8.3). The technique was primitive and the doses given to healthy tissues too high, so that in 1948 Stone evaluated the effects of neutron therapy on 226 patients and concluded: "*Neutron therapy as administered by us has resulted in such bad late sequels in proportion to the few good results that it should not be continued.*"

The oncologist and radiotherapist Mary Catterall understood that the technique had been improperly used and in 1965 revived neutron therapy at Hammersmith Hospital in London; later fast neutron centres were built in many countries as a

---

[3] A deuteron, which is made of a proton and a neutron strongly bound by the nuclear force, is an isotope of hydrogen. Hitting a nucleus of the target, the deuteron breaks up and the neutron is projected forward, becoming what physicists call a *fast* neutron.

consequence of the first enthusiastic reports from Hammersmith. The worldwide effort in neutron therapy has been large; however, at present, this technique is rarely used even though it has been demonstrated that it can control radioresistant tumours of the salivary glands.

The biological effects of neutrons are due to secondary protons set in motion when neutrons traverse tissues and collide with hydrogen nuclei. These secondary low energy protons are 'densely ionizing' particles, because they produce ionizations that are closer together than the diameter of the DNA double helix; this property was quickly understood to be the main reason for their effectiveness in killing radioresistant cells.

Neutrons have been substituted by carbon ion beams, which are a better alternative to deliver doses with densely ionizing particles without the deposit of large doses in *all* the traversed tissues, which is typical of X-rays (Fig. 7.7 on p. 198) and also of fast neutrons.

Radiotherapy with Protons and Light Ions: Wilson Proposal

Neutron therapy is one of the many types of treatments that use beams of 'heavy' particles. These particles are made of quarks and, since such composite particles are called 'hadrons', the technique is often called 'hadron therapy'.[4] Others, to distinguish it from X-ray radiotherapy, prefer the terms 'ion beam therapy' or 'particle therapy', which however is a misnomer since photons of an X-ray beam are also 'particles', as lengthily discussed in the first chapters. Another very much used misnomer is 'heavy ion therapy', since for physicists carbon, oxygen and similar ions are 'light', the name 'heavy' being reserved for ions larger than iron.

Today only two types of hadron are used to treat solid tumours – *protons* and *carbon ions* – but attempts have been made, and abandoned, to use neutrons, as discussed above, and also beams of helium, neon, silicon and argon ions, and also of charged pions, the unstable particles first discovered in cosmic rays (Figs. 1.16 on p. 32 and 4.1 on p. 95). Protons and carbon ions are bare nuclei obtained by removing, for example with an electric discharge, the single electron from a hydrogen atom or all six electrons from carbon atoms, which remain nuclei composed of 6 protons and 6 neutrons.

---

[4] I introduced the term 'hadrontherapy' in 1992 when I wanted to coin a collective term that would include all types of non-conventional radiation beams used at the time: protons, helium ions, neon ions, neutrons and pions. Indeed physicists call 'hadrons' all the particles that feel the strong interaction because they are made of quarks and antiquarks. The name sounds fine in English, French (hadronthérapie), German (Hadronentherapie), Italian (adroterapia) and Spanish (hadronterapia). In English, as initially proposed, I prefer to write 'hadrontherapy' (or 'hadrotherapy') as a single word because 'radiotherapy' was initially written as two words, which were joined together when the technique became a standard medical procedure.

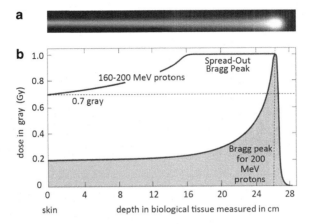

Fig. 8.4 (**a**) Penetrating into a biological tissue sample, a narrow mono-energetic proton beam produces the largest number of ionisations at the end of its path, in a well-defined 'spot' which has a 10 mm diameter (Courtesy of Paul Scherrer Institut, Center for Proton Therapy, Villigen). (**b**) Numerous superimposed Bragg peaks, caused by a beam containing different numbers of protons with energies in the range from 160 to 200 MeV, give a uniform dose to a tumour of length 10 cm

The history of this technique began in 1946 when Bob Wilson was called to lead the team for design and construction of a new 160 MeV cyclotron in Harvard, where he had been appointed associate professor after leaving Los Alamos. He spent one year in Berkeley, collaborating with Ernest Lawrence, who had been his professor in the early 1930s, to complete the design of the accelerator. It was then that Lawrence asked him to define the shielding of the new cyclotron, by calculating the interactions with matter of a beam of 100 MeV protons. Wilson followed his suggestion and found that the proton dose has a completely different trend with depth than a beam of X-rays.

Suppose we bombard skin with a narrow beam of protons, all of the same energy. As they pass through the tissue, the protons remove electrons from molecules, ionising them while slowing down. The maximum number of ionisations per millimetre is found at the end of the range in matter and is called 'the Bragg peak', after the English physicist William Bragg who was first to discover its existence for alpha particles; the peak occurs at a depth that depends on proton energy. In the example of Fig. 8.4, with protons of 200 MeV a very pronounced Bragg peak is found at 26 cm depth in soft tissue.

By simultaneously using protons of different energies many peaks can be superimposed, which form a 'Spread-Out Bragg Peak' with which a target tumour of any length at any depth can be fully exposed.

In the case of Fig. 8.4 the dose to the skin is limited to 60 % of the maximum, which is what is delivered to the tumour. Moreover the normal tissue beyond the target tumour does not receive any dose, because the protons stop; this is very different from what happens with X-rays, which cross the entire body of the patient (Fig. 7.7 on p. 198).

Fig. 8.5 (**a**) Bob Wilson (*centre*) in 1943 at Harvard, accompanied by the radiation therapist Hymer Friedell, on the *left*; on the *right* Percy Bridgman (Courtesy Harvard University, Collection of Historical Scientific Instruments). (**b**) Wilson in front of two panels of the exhibition "Atoms for health" displayed in 1996 during the CERN Symposium (Courtesy CERN)

These observations prompted Bob Wilson to propose the use of protons for irradiating solid tumours, as a better therapy than the one based on X-rays. His pioneering and now famous paper – "Radiological Use of Fast Protons" – was published in 1946 in the journal *Radiology*. It is interesting to remark that in his work Wilson discussed mainly protons but he also mentioned alpha particles and carbon ions with the following words:

> The intense specific ionization of alpha particles will probably make them the most desirable therapeutically when such large alpha particle energies are attained. For a given range, the straggling and the angular spread of alpha particles will be one-half as much as for protons. Heavier nuclei, such as very energetic carbon atoms, may eventually become therapeutically practical. (Wilson 1946)

The planning of the 160 MeV cyclotron in Harvard, aimed at nuclear physics experiments, had thus opened the way to a new important medical application. Harvard's facility accelerated its first proton beam in 1949 and only after many years of exploitation for fundamental research, would it be used in 1961 to irradiate patients, as we will see in the next paragraph.

Two pictures of Wilson, taken more than 50 years apart, are shown in Fig. 8.5. The first photo was taken in 1943 in interesting circumstances. Harvard University had built its first cyclotron in 1937, using drawings obtained from Lawrence, but the federal government drafted it during World War II. It was taken apart and shipped to Los Alamos in 1943, for service in designing the first atomic bombs. The move had to be made in secrecy, so the young Bob Wilson, together with the radiation therapist Hymer Friedell, was sent from Los Alamos to Harvard to negotiate the purchase and arrange the transfer; the cover story was that the cyclotron had to be moved for medical purposes, the treatment of military personnel. Percy Bridgman, then physics department chairman and future Nobel Prize winner in Physics in 1946, told the government agents they could not have the machine if they were going to use it for medical purposes. But, he added, "*If you are going to use it for what I think you're going to use it for, you are welcome to it*" (Wilson 2004).

The second picture was taken in 1996, 50 years after the publication of the *Radiology* paper. CERN with the Swiss Paul Scherrer Institute (PSI) were organizing the 'Second International Symposium on Hadrontherapy' and Bob Wilson was 82 and not in good health. Thus, to convince him to travel to Europe, instead of writing I called him at his home in the United States. He had not noticed that his proposal was exactly 50 years old, but immediately accepted to come and give the opening speech.

This was a moving event in the overcrowded CERN auditorium and a beautiful way to celebrate his pioneering work. A few weeks later I transcribed the speech from the recordings and, after some minor editing, I sent it to him for corrections before publishing the proceedings of the symposium. I never got an answer and so the text appeared uncorrected by him; this had been his last public speech. I quote an excerpt.

> *I did have something of a conscience that was somewhat out of joint. At Los Alamos, where I had been working for the past five years, no matter how justifiable it may have been, we had been working on one thing, and that was to kill people. When that became crystallized in my mind by the use of the atomic bomb at Hiroshima, it was a temptation, to salvage what was left of my conscience, I suppose, and think about saving people instead of killing them.*
>
> *I jumped into the most obvious thing I could do next: because one could hurt people with protons, one could probably help them too. So I tried to work out every detail and I was surprised to see, when all the details were taken into account that the Bragg curve came up and came down very sharply. Another thing I found was that the protons followed an almost straight line, with just a little wiggle at the end, so one could deposit a large fraction of the energy into a few cubic centimetres. I thought all that was great and thus I sent the paper to Radiology.*
>
> *How was this received by the radiologists at the time? I must say it was received as though a lead balloon had been dropped in their midst, not at all in a friendly way. Either it was ignored or the people did not like it at all. So it has taken quite a while in these fifty intervening years before one can say that the paper was received.* (Wilson 1997)

First Treatments in Berkeley, Uppsala and Boston

Two years after Wilson's paper, researchers at the Berkeley Laboratory conducted extensive studies on proton beams and confirmed his predictions. After many animal irradiations, the first patient was treated in 1954 under the guidance of Cornelius Tobias, a Hungarian physicist who arrived in 1939 at Berkeley, aged 19, with a Hungarian-American Fellowship. Toby, as everybody called him for the rest of his life, immediately saw Ernest Lawrence and told him of his interest in possible applications of nuclear physics to medicine. Sixty years later he described this encounter.

> *His blue eyes were earnestly looking at me over through gold-rimmed glasses. 'I have a long-standing interest in medicine' he said 'and my wife Molly has a degree in microbiology. My brother John is a physician and I almost became one myself. The problem with your goal is that there is hardly any individual alive who can understand both physics and biology. How do you plan to acquire the necessary knowledge?'* (Tobias and Tobias 1997)

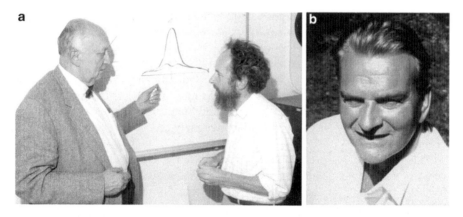

Fig. 8.6 (**a**) In 1986 in Berkeley Cornelius Tobias (on the *left*) discusses the transverse distribution of the ion dose and the relative effectiveness, inside a single ion track, with Gerhard Kraft, a German nuclear physicist and biophysicist who spent one year in Berkeley and later introduced ion therapy in Germany (Courtesy Lawrence Berkeley National Laboratory). (**b**) Börje Larsson, who, in 1957, irradiated the first human tumours with a proton beam (Courtesy of Inga-Lis Larsson)

They decided that Toby would first get a standard master's degree and a doctorate in nuclear physics; thus he learned accelerator physics from Lawrence himself, Oppenheimer, Alvarez, Segrè and many others. From 1942 he worked as a 'medical physicist' with John Lawrence to treat leukaemia with radioactive phosphorus.

In time Tobias became an internationally known radiobiologist and led a very large group of scientists working for decades at the frontiers between physics, biology and medicine. No surprise then that in 1954 he was, with John Lawrence, behind the first hadron treatments on humans: the irradiation of the pituitary gland in patients with metastatic breast cancer that was hormone sensitive. The rationale was that proton treatment would stop the pituitary gland from making hormones that stimulated the cancer cells to grow. The pituitary was a natural site for the first treatments, because the gland location was easily identified on standard X-ray films and the breast cancer could be removed surgically. Between 1954 and 1974 about 1,000 pituitary glands and pituitary tumours were treated with protons with a 50 % success rate.

In 1957 the first tumour was irradiated with protons at the cyclotron of the Gustaf Werner Institute in Uppsala by Börje Larsson (Fig. 8.6b), who obtained his PhD in 1962 discussing the subject: "Application of a 185 MeV Proton Beam to Experimental Cancer Therapy and Neurosurgery: a Biophysical Study". Many years later Larsson told me that during those years he was very much inspired by Tobias, who in 1954 spent a sabbatical year in Uppsala.[5]

---

[5] Börje Larsson was the first person I went to visit in Zurich when, in 1991, I became interested in proton and 'light' ion therapy. We became good friends and in 1993 we organized a symposium in Como, which was attended by all the important scientists working in the field. In 1994 we edited

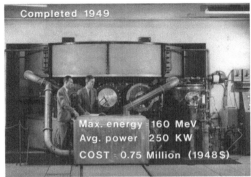

Fig. 8.7 (**a**) Photograph of the brand new 160 MeV Harvard cyclotron (Courtesy Harvard University, Collection of Historical Scientific Instruments). (**b**) Herman Suit (*right*) and Michael Goitein in 2003, when Goitein received the Medal of the American Society for Radiation Oncology (Courtesy of George TY Chen)

The facility that made the largest impact on the development of proton therapy is certainly the 160 MeV Harvard cyclotron (Fig. 8.7a), in Boston.

This was the cyclotron at which Bob Wilson was working in 1946 and was undoubtedly the best place to start proton therapy. In 1961 – many years later – Raymond Kjellberg, a young neurosurgeon at Massachusetts General Hospital (MGH) in Boston, became the first to use the Harvard beam to treat a malignant brain tumour. In the following years the story of Harvard's cyclotron became complicated, reflecting on a minor scale the very long and tortuous path which proton therapy took to become established, as described in 1999 – at the 50 year anniversary ceremony – by Richard Wilson, who was for many years one of the main promoters of the use of the cyclotron in cancer therapy.

Already in 1965 Dr Kjellberg discussed with Massachusetts General Hospital whether they could take over operation of the laboratory. There followed a five-year period of discussion, irrevocable decisions, revoking the decisions, and further discussions. All options were considered. Moving the cyclotron to another university (such as Northeastern) who wanted it; allowing another organization, hospital, university or merely different faculty, to run it; or closing completely.

A preferred alternative to dismantling the cyclotron was to 'give' the cyclotron to Massachusetts General Hospital or Harvard Medical School. Negotiations continued about this throughout the next years. But physicians thought that the resources of the school were better committed to finding the causes of cancer than of treating it. Moreover, cancer experts were arguing that chemotherapy was a more promising choice for patient treatment than radiotherapy. While both these arguments seemed plausible at the time, it is now clear that they were wildly overoptimistic. Thirty-five years later the causes of cancer are still elusive, and chemotherapy, by itself, is far less effective (for solid tumours) than when combined with radiotherapy.

---

the proceedings, 'Hadron therapy in Oncology', which is the best source of information on the status of ion beam therapy at a turning point of its long history.

The cyclotron was finally rescued and remained in Harvard by two important steps. Andreas Koehler (at the time the physicist Andy Koehler was the only part time employee working at the cyclotron) proposed a budget to the physics department showing that the cyclotron could be kept alive for one or two days a week, funded by patient fees from Dr Kjellberg's patients. The second step was the arrival in Boston of Dr Herman Suit, to become the new chief of the newly established Department of Radiation Medicine at MGH. One of Dr Suit's first appointments was of the physicist Dr Michael Goitein, who was an expert in the use of computers. The stage was now set for a most productive thirty year period of the operation of the Harvard Cyclotron. (Wilson 2004)

By the mid-1970s at the Harvard Cyclotron Laboratory the physicists Andy Koehler, Bernard Gottschalk and their colleagues – working with radiation oncologists guided by Herman Suit – had developed methods to treat large brain tumours, while Michael Goitein had written very sophisticated codes for quantifying the related treatment plans. The medical successes impressed the American National Institute of Health enough to provide funds to rescue the machine.

Overall three groups of radiation oncologists worked for many decades with Harvard physicists on three clinical studies: neurosurgery for intracranial lesions (3,687 patients), eye tumours (2,979 patients) and head-neck tumours (2,449 patients).

The main people who did work on eye tumours and malformations were Ian Constable and Evangelos Gragoudas of the *Massachusetts Eye and Ear Hospital,* who used hardware and software tools developed by Koehler and Goitein. Since for eye therapy protons have to penetrate only 2.5–3 cm of tissue, their maximum energy is 60–70 MeV. Following the Harvard results, many hospitals have built centres based on relatively small cyclotrons that provide such energies. By 2012 about 15,000 patients had been irradiated at these facilities with a cure rate greater than 95 % and with about 80 % of the patients retaining their sight.

The results obtained convinced many radiation oncologists of the superiority of protons with respect to X-rays to treat tumours that are close to organs at risk. In particular, for two rare tumours at the base of the skull, chordomas and chondrosarcomas, the *local control rate* at 5 years was 80 % and 95 % respectively, while it was about 40 % with X-rays. (The local control rate is the fraction of patients who, after 5 years, do not show any sign of regrowth of the treated tumour. For many tumours the survival rate is smaller than the control rate because in a number of cases the tumour has meanwhile metastasised and the patient does not survive).

By 2002, when the cyclotron was finally closed, almost 10,000 patients had been irradiated and foundations laid for the subsequent rapid development of the field. In particular, the expertise developed in Harvard was soon transferred to the new 230 MeV hospital-based facility of the Massachusetts General Hospital, which was built by the Belgian company Ion Beam Accelerators (IBA) and opened in 2001.

The Precision and Efficacy of Protons

Nowadays precision radiology gives detailed pictures of tissues and their metabolism and allows the radiation oncologist to define accurately the tumour target and the critical organs to be avoided. For about twenty years, to best avoid organs at risk and concentrate the dose on the target, many crossed beams of non-uniform intensity used to be employed in what is called Intensity Modulated Radiation Therapy (IMRT). In more recent years this technique has been improved by detecting in real time the position of the tumour, which may move during the irradiation because of patient respiration, and to follow it by changing continuously the direction of the X-ray beam in what is called Image Guided Radiation Therapy (IGRT). This technique ensures coverage of the tumour, even when it moves.

Figure 8.8 shows, in the case of a brain tumour, a comparison between the dose distributions due to nine X-rays beams and four proton beams. The colours quantify the doses received by the different tissues according to the scale reproduced at the left of the figure.

It is clear that protons spare organs at risk better than X-rays, even if these are applied with the most modern techniques (Intensity Modulated Radiation Therapy).

The dose saving which is achieved with protons is especially crucial for children; their tissues, still in the development phase, are significantly more sensitive to the damaging effects of radiation. The reduction in the dose given to these tissues not only permits balanced growth of the organs (and, in the case of cerebral irradiation, a reduction of frequent cognitive impairment) but it should considerably reduce the likelihood of growth of a secondary tumour in later life – a beneficial effect that is difficult to pinpoint. For these reasons, in the USA the number of children treated with protons exceeded one thousand in 2013, increasing by about 10 % per year.

In all irradiations with protons the greater concentration of dose on the target tumour permits an increase of dose given to diseased tissue without adding to the damage to healthy tissue, in comparison with the best X-ray methods. This is a vital contribution to improved treatment because even a small addition to the dose given to cancerous cells has beneficial consequences. For example, let us consider a certain type of tumour, which 5 years after the first therapy cycle is controlled in 50 % of cases with a specific X-ray dose. If a proton dose just 10 % larger can be delivered, the probability of controlling it typically increases from 50 % to 65 %; put in another way, the probability of a reoccurrence of the cancer is reduced from 50 % to 35 %.

In summary, with *proton beams* the *same* dose as with X-rays can be delivered to cancer cells, producing in them the *same biological and clinical effects*, but giving a lower dose to surrounding tissue; this advantage can be exploited by increasing the dose to the tumour, and thus the probability of controlling it, while having fewer negative consequences for the healthy tissues.

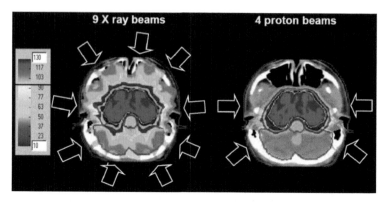

Fig. 8.8 A comparison between two treatment plans with X-rays and protons which maximize the doses given to a brain tumour by minimizing the doses deposited in the healthy tissues (Courtesy of Paul Scherrer Institut, Center for Proton Therapy, Villigen)

From Physics Laboratories to Hospital Based Centres

In the years following the start-up of the Harvard cyclotron as a medical facility, in 1961, other nuclear physics laboratories in USSR and Japan assigned easy-to-build horizontal beams of their research facilities to proton therapy (Table 8.1). Eventually in 1984 the Paul Scherrer Institute (PSI), in Villigen – Switzerland, entered the field becoming the leader in developing irradiation techniques for particles to enter the patient's body from any angle.

Looking back, one may ask why it has taken almost 50 years for *proton therapy* to become a well-recognised and widespread methodology; indeed, it was necessary to await the end of the 1980s for the number of patients treated to exceed a threshold of 10,000. The reasons are easy to explain.

For X-ray therapy of deep tumours, it is enough to accelerate electrons up to 10 MeV; instead protons must be taken to at least 200 MeV, an energy 20 times higher. Moreover, the proton mass (1 GeV, i.e. 1,000 MeV) is 2,000 times larger than that of the electron (0.5 MeV); medical proton accelerators – cyclotrons or synchrotrons – are therefore much larger and more expensive than linacs which produce X-rays. For these reasons, until the beginning of the 1990s, accelerators normally used to carry out nuclear and particle physics experiments were employed – often part-time – to treat tumours in conditions which were far from ideal. In this respect PSI was an exception.

A smooth transition from a physics laboratory to a hospital facility took place in Japan. The University of Tsukuba started proton clinical studies in 1983 using a synchrotron constructed for physics studies at the High Energy Accelerator Research Organization (KEK). About 700 patients were treated at this facility from 1983 to 2000. In 2000, a new in-house facility, called the *Proton Medical*

Table 8.1 The pioneers of proton therapy; these facilities are now no longer in operation except for PSI and the centres in Moscow and St. Petersburg

Facility	Country	Years of operation
Lawrence Berkeley Laboratory	USA	1954–1957
Uppsala	Sweden	1957–1976
Harvard Cyclotron Laboratory	USA	1961–2002
Dubna	Russia	1967–1996
Moscow	Russia	1969–now
St. Petersburg	Russia	1975–now
Chiba	Japan	1979–2002
Tsukuba	Japan	1983–2000
Paul Scherrer Institute (PSI)	Switzerland	1984–now

Research Centre (PMRC), was constructed adjacent to the University Hospital. Built by the Hitachi company, it was equipped with a 250 MeV synchrotron.

However, the first hospital-based proton therapy centre was built at the Loma Linda University Medical Centre (LLUMC, California), as a consequence of the long-standing determination of the radiation oncologist James Slater, who convinced Fermilab management to design and build a 7 m diameter synchrotron which would accelerate protons up to 250 MeV. The first patient was irradiated in 1990 and, in the same year, USA Medicare approved the insurance coverage of this new modality for treating cancer. By 2013 more than 20,000 patients had been treated at LLUMC.

It was not by chance that Fermilab built the synchrotron. The roots of the collaboration with the LLUMC are to be found in 1972, when Bob Wilson, then Fermilab director, wanted to use the proton beams extracted from one of the accelerators for tumour therapy. However, use of the proton Bragg peak requires precise knowledge of the densities of tissues and the location of the tumour and the technology to perform this localization was not fully developed at that time; thus Fermilab concentrated on neutron therapy.

After Wilson resigned in 1978, the new director Leon Lederman – who, 10 years later, shared Nobel Prize in physics for the discovery of the second neutrino (Fig. 4.13 on p. 114) – agreed to the Slater proposal and the design of a 230 MeV synchrotron started in 1986. Meanwhile the problems related to localization of the spread-out Bragg peak in the tumour had been overcome by new imaging modalities, in particular computerized tomography (CT), and by treatment planning systems developed by Goitein. The subsequent development of MRI,[6] SPECT, and PET scanning has further improved the target definition, allowing millimetre precision of dose delivery with hadron beams.

---

[6] An apparatus for magnetic resonance imaging (MRI) does not make use of particle accelerators, but takes advantage of superconducting solenoids (at the centre of which slides the couch carrying the patient) which are derivatives of the superconducting magnet technology used by large particle detectors and accelerators.

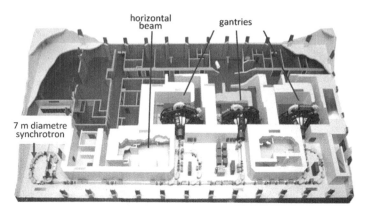

Fig. 8.9 LLUMC features a 7 m diameter synchrotron, built by Fermilab, a horizontal beam and three rotating gantries, which have a 10 m diameter and support a system of bending magnets and quadrupoles allowing irradiation of the patient from any direction (Courtesy of Loma Linda University Medical Center)

LLUMC construction has been fundamental to the progress of proton therapy through its pioneering realizations of the rotating gantry, which allows adjustment of the angle of the beam penetrating into the patient's body (Fig. 8.9). An important factor for an oncological facility is the ability to treat as many patients as possible and each in a short time. For this reason patients are irradiated at LLUMC in four different treatment rooms; one has a horizontal beam and the other three are equipped with rotating heads, so called 'gantries'. In this way, while a patient is irradiated for 3–5 min in one room, in others precise alignment measurements are taken to prepare other patients. Because each treatment requires 20–30 sessions, one centre with four rooms can manage up to 1,200–1,300 patients a year.

Twenty years after the beginning of operation of Loma Linda University Medical Centre, there were almost 40 proton therapy centres in operation or under construction throughout the world. Seven companies offer turnkey proton therapy centres, organised around a cyclotron – either room temperature or superconducting – or a synchrotron of 6–8 m diameter. To fully exploit the accelerator, most hospitals have from three to five treatment rooms; however since 2010 'single-room' facilities are offered to customers who cannot afford initial investments larger than 100 million euros and aim at treating no more than 300 patients per year.

Figure 8.10 shows a multi-room centre of the Belgian firm Ion Beam Applications (IBA), market leader in proton therapy facilities: typically it features three gantries, each one with a diameter of 10 m and a weight of almost 100 t.

IBA was founded in 1986 by the Belgian engineer Yves Jongen (Fig. 8.11). He won the 2013 European Inventor Award; in a broadcast on this occasion he said: "*I spent hundreds of hours on my knees in this machine changing the parts to try to make it perform a bit better, to make something fully reliable. This machine is therefore truly my baby.*" (Jongen 2013)

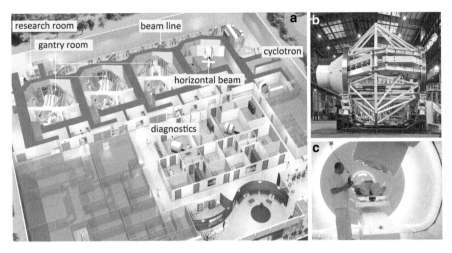

Fig. 8.10 (**a**) A multi-room centre for ion beam therapy designed by the company IBA. (**b**) Magnets mounted on three rotating gantries deflect the proton beam onto the patient lying on the couch. (**c**) By rotating the gantry the beam can be directed along the directions that minimise the dose deposited in the organs at risk (Courtesy of IBA)

Fig. 8.11 In 1995, during a visit to the IBA premises, Yves Jongen (*right*) shows a pole of a 230 MeV cyclotron to the author and to Sandro Rossi (*left*), then Technical director of the TERA Foundation (Ugo Amaldi)

At the end of 2013 more than 100,000 patients had been treated in the world with proton therapy, a number that grows exponentially at a rate of about 8 % each year.

The Precision and Efficacy of Carbon Ion Therapy

An atom of carbon – made of six electrons that rotate around a nucleus made in turn of six protons and six neutrons – becomes a 'carbon ion', as used in radiotherapy, when it is deprived of its six electrons. The ionization produced by a beam of carbon ions – all having the same energy and stopping in tissue – causes a Bragg spot similar to that of Fig. 8.4a on p. 223, but at a 200 mm depth its width in tissue is 3–4 mm instead of 10 mm. The dose distribution for the brain tumour of Fig. 8.8 is similar to those of protons, but has the advantage of sharper edges because of the reduced diameter of the spot. This is clinically important to spare organs at risk, which are close to the irradiated target, but for a non-expert eye it is difficult to distinguish on a picture, as in the one of Fig. 8.8, the dose distribution resulting from irradiation with carbon beams from one obtained with protons.

If carbon ions and protons both spare the organs at risk, why are carbon ions used at all? The first reason has already been mentioned; the delivered doses have sharper edges than in the proton case. The second, more important reason is related to the distribution of the ionizations along the particle path in tissue.

As shown in Fig. 8.12, for ions the spatial distribution of the DNA lesions is characterized by precise tracks, unlike the sparse and distributed ionization pattern generated by X-rays.

At the end of these tracks, before stopping, carbon ions directly produce double-strand breaks, which are often clustered, and thus mostly irreparable. This is the second advantage of carbon ions with respect to protons, and also to X-rays.

Why in this respect are carbon ions and protons different? Carbon ions, having electric charges six times larger than protons, slow down more rapidly; indeed, to reach 300 mm depth a carbon ion must have initially about 5,000 MeV energy, instead of about 200 MeV for a proton. In the slowing down process they detach many more molecular electrons from the tissues they cross: on average the number of ionisations produced by a carbon ion is $5,000/200 = 25$ times that caused by a proton which reaches the same depth. The consequence is that, if a beam of carbon ions deposits the same dose in tumour cells, *at the molecular level* the biological effects are *different* from those of protons.

In the last few centimetres before stopping a proton is *sparsely* ionizing while a carbon ion is *densely* ionizing and produces many nearby ionizations, the distance between two consecutive ones being, on average, smaller than two nanometres, the transverse dimension of the double helix. If more than one ionisation is caused in such a short segment of trajectory, the DNA is directly damaged. Thus, while protons (and X-rays) damage DNA essentially by means of the *indirect* effects of free radicals, carbon ions at the end of their path in tissue have *direct* effects on the DNA, producing double strand breaks (DSB) and, in particular, clustered damages

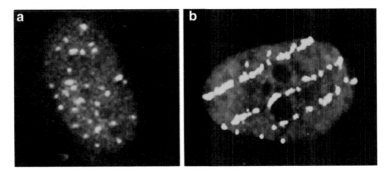

Fig. 8.12 Nuclei of human fibroblasts, cells of connective tissue, have been irradiated with X rays and ions. The green points are double-strand breaks (DSB) detected with a special staining technique. (**a**) With X-rays the DSBs are uniformly distributed in a nucleus. (**b**) The DSBs of ions are found along three well-defined tracks (Cucinotta and Durante (2006))

so severe that the cell can hardly repair them. Moreover, because the direct effect is now the main pathway, the oxygen enhancement ratio (OER) is reduced and the therapy is more effective against hypoxic tumours.

In short, from the biological and clinical point of view, protons and X-rays are similar to each other because they are *sparsely ionizing* radiations and 80 % of the killed cells are killed by *indirect* effects.[7] A beam of carbon ions is by its nature a very different radiation, which macroscopically has roughly the same overall dose distributions as protons but in the tumour target, where the ions stop, is *densely ionizing* and in almost 80 % of the cases induces cell death by *direct* effects, which are not sensitive to the presence of oxygen. This does not apply in the tissues close to the skin, where carbon ions also behave as sparsely ionizing radiation and have less damaging effects.

This different mechanism makes carbon ions capable of treating radioresistant and hypoxic tumours, which represent about 5 % of the 2,000 tumours irradiated each year with X-rays in a population of one million inhabitants, and possibly many more.

The road to the treatment of radioresistant tumours with carbon ions has been a very long one in spite of the fact that Bob Wilson proposed it in his seminal paper already in 1946. Thirty years later at Lawrence Berkeley Laboratory, the group of radiobiologists led by Cornelius Tobias started to develop ion beam hadron therapy, initially experimenting with cell cultures and animals. In 1975 patients were treated for the first time at the Bevalac[8] accelerator with argon ions, which are made up of

---

[7] In the last millimetres of their path in tissue, slow protons are also densely ionizing and give some distinctive effects compared to photons.

[8] The Bevalac was built at Berkeley in 1974 by connecting the recent super-HILAC (Heavy Ion Linear Accelerator) to the old proton synchrotron Bevatron with a 300 m long transfer line made of many bending magnets and quadrupoles.

18 protons and 18 neutrons and seemed good candidates against hypoxic and otherwise radioresistant tumours. But problems arose owing to non-tolerable side effects in the normal tissues. After a few irradiations of some 20 patients, Tobias and collaborators used lighter ions, first silicon for two patients and then neon ions – which are made of ten protons and ten neutrons – for 433 patients, until the Department of Energy decided to close the Bevalac.

At the closure of the programme, in 1993, the results obtained led to the conclusion that neon ions had too high an electric charge and damaged the healthy tissue through which the beam passed. In contrast, carbon ions, having only six protons, were optimal for the treatment of radioresistant tumours, as well as for the preservation of healthy tissue.

Two Protagonists of Carbon Ion Therapy Outside the United States

In the 20 years following the closure of the Berkeley ion programme nothing has happened in the United States, while carbon ion therapy has been greatly developed in other continents; as a consequence of the construction of dedicated centres in Japan, Germany and Italy in 2013 the total number of patients treated throughout the world has passed 10,000. In Japan and Germany the main characters of these successful stories have been Hirohiko Tsujii and Gerhard Kraft.

Tsujii is a medical doctor and radiation oncologist, who in the 1980s treated patients with protons at the Tsukuba centre and with pions at Los Alamos. Afterwards he was called to NIRS, the Japanese National Institute for Radiation Research, to lead the medical group which in 1994 started irradiating patients with beams of carbon ions at the synchrotron HIMAC (*Heavy Ion Medical Accelerator in Chiba*), built in the Chiba Prefecture about 30 km from Tokyo (Fig. 8.13).

This institute, world leader in carbon ion therapy, treated over 8,000 patients by the end of 2013. The results have been so promising, including the treatment of common cancers such as tumours of the lung, liver, pancreas and local recurrence of rectal cancer, that three other centres for carbon ions and protons have been built in Japan.

In 2007 Hirohiko Tsujii gave an interview to the CERN Courier:

> *The reason we decided to use carbon ions rather than protons is that it is the most balanced particle: it has the property of providing a uniform treatment to the tumour and also has a higher biological effect on the tumour. It took several years before coming to the optimum level of treatment with carbon ions. Now the local control for almost all types of tumours is 80–90 %, and – after choosing the optimal level of treatment – the local control is expected to be more than 90 %. Another point that I want to focus on is the use of 'hypofractionated' radiotherapy. Stage I lung cancer, cervical and prostrate cancer or other large tumours require only 16–20 fractions against around 40 fractions using conventional treatments.*
> (Lee 2007)

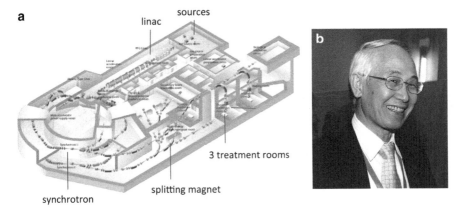

Fig. 8.13 (**a**) HIMAC features two large synchrotrons, injected by an Alvarez linac, and three treatment rooms for a total of two horizontal and two vertical beams (National Institute for Radiological Science, Japan). (**b**) Hirohiko Tsujii has led the clinical programme since 1992 to 2012 (Courtesy of CERN)

The nuclear physicists and radiobiologist Gerhard Kraft devoted himself for many years to the realization of the German pilot project for oncological ion therapy. After getting his PhD at the University of Cologne, he had spent two sabbatical years with Cornelius Tobias in Berkeley (Fig. 8.6a on p. 226). His experience in the United States inspired his subsequent research activities. Back in Germany, he became a staff member of the German nuclear physics laboratory GSI (*Gesellschaft für Schwerionenforschung*) near Darmstadt, where heavy ions were accelerated and used in fundamental nuclear physics experiments. There he started a vigorous radiobiological programme on cells and proposed to create a beam of 5,000 MeV carbon ions for treating patients. The 'pilot' facility was eventually built with its treatment room featuring an horizontal beam, but the approval process was very long, as he later described:

> *In 1988 a proposal was submitted to the German Government but never acted upon by the Ministry. This first proposal was followed by later attempts to obtain funding from national and international sources. These failed too, partially because of the unusual structure of the project with a very artificial separation between beam production by GSI on one side and all the other being the responsibility of the University Radiology Department and of the Heidelberg cancer research centre (DKFZ). At that time GSI did not want to be involved in the project except for beam production. The great fear was that ion therapy would be too successful and could overshadow the nuclear and atomic physics programs at GSI. This view changed completely in the spring of 1993 when Hans Specht took over the directorship of GSI. He wanted to see the first patient to be treated within the first four years of his directorship.* (Kraft 2013)

The treatment at the pilot facility started in 1997 and novel technologies were developed for depositing the dose in the tumour target and monitor it. By 2008 about 450 patients were irradiated under the direction of the radiation oncologist Jürgen Debus, who is a medical doctor with a physics PhD. In 2009 the pilot project

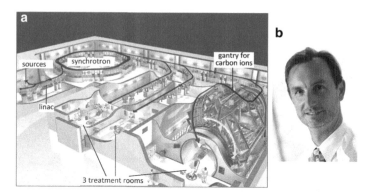

Fig. 8.14 (a) The Heidelberg Ion-Beam Therapy Centre (HIT) features two horizontal beams and a carbon ion gantry, the first one ever built (Courtesy HIT). (b) Jürgen Debus, who is a physician with a PhD in physics, has been the DKFZ radiation oncologist behind both the GSI pilot project and HIT (Courtesy HIT)

was closed and the brand new Heidelberg Ion-Beam Therapy Centre (HIT) was opened, retaining all the medical and technical competence accumulated in Darmstadt. The HIT was a joint endeavour of the GSI and the Siemens Medical company.

A carbon ion gantry needs more powerful and weightier magnets than a proton gantry because the trajectories of 5,000 MeV carbon ions are three times more difficult to bend than beams of 200 MeV protons. Thus the HIT gantry of Fig. 8.14 is 25 m long and weighs 600 t, while a typical proton gantry is half the length and at least six times lighter; furthermore, HIT's gantry can also be used with protons. HIT has thus been the first hadron therapy centre that could compare clinical results obtained with protons and carbon ions, with beams coming from the optimal direction to avoid organs at risk.

A Personal Journey

The beginning of the 1990s signalled a period of change in my professional life, a period in which various positive events encouraged me to reconsider the future of my scientific work and decide on a change of course. At the international particle physics conference held in Singapore in August 1990 I presented – on behalf of the collaboration of which I was spokesman – the results we had obtained at LEP in one year of data taking with the DELPHI experiment (Fig. 5.3 on p. 128). In the same crowded session the spokespersons of the other three LEP collaborations – ALEPH, L3 and OPAL – also reported and those of us from DELPHI learned with great satisfaction that we had been the first to measure the three-gluon coupling (Fig. 5.11b on p. 139), a quantity whose value is fundamental in the Standard Model.

To this team success was soon added a more personal achievement. In the same period, I realised that the measurement of the strong coupling constant obtained in the first year of DELPHI data taking had reached a precision sufficient to allow an analysis using experimental data of the unification of the strong and electro-weak forces. Working for a few weeks with my friend Wim de Boer and his PhD student Hermann Furstenau, as I described earlier in Chap. 6, we obtained the results illustrated in Figs. 6.16 and 6.17 (p. 175); introducing the simplest supersym-metric model we were able to demonstrate that unification of the strong, weak and electromagnetic couplings should occur at very high energy.

At the beginning of 1991 I recognised how difficult it would be for me to achieve something more significant in the field of particle physics. I was 57 years old and for some years I had taught a postgraduate course at the University of Milan in radiation physics, which had led me to discover the most recent applications of physics to cancer treatment. It had been 20 years since I had left the National Health Institute where I had begun my research career in Rome. For the radiation physics course I reviewed much of the literature – reading, among other things, the studies of treatment using light ions carried out at Berkeley – and discussed them with a long-standing friend, Gianpiero Tosi, the most noted Italian medical physicist, who taught in the same school and had independently developed the same interests. So, in May 1991, we wrote a report together entitled "A future hadron radiotherapy centre" in which we proposed to design a hospital therapy facility using light ions and protons which should be constructed in Italy.

In August I discussed our project with Nicola Cabibbo, one of the great Italian theoretical physicists, who was then President of the Italian National Institute for Nuclear Physics (INFN) and who immediately encouraged me to go ahead with a funding request. In October the first financial allocations were made, to organise meetings among the leaders in hadron therapy in Europe and the USA.

In the meantime I had accepted a chair in particle physics at the University of Florence, which was then located in the hills of Arcetri, the place where Galileo had worked and taught. During some months of the 1991–92 academic year I spent 2 or 3 days each week in that extraordinary place, teaching an enthusiastic group of youngsters the physics which excited me. This experience provided the time and detachment necessary to decide my future; I would dedicate myself to cancer therapy with hadron beams. Shortly afterwards I resigned as DELPHI spokesman and at the end of 1993 I left that rewarding task after 12 years.

My wife sometimes says that in this way I returned to my first love – that of the 1960s in the Rome National Health Institute – and there is some reproof in her voice for the time and energy which I have devoted to particle and medical physics. I know this comes from affection, but each time I regretfully acknowledge that these two activities have much absorbed me, perhaps too greatly, taking time from our four children and seven grandchildren; it is not a simple compromise to divide one's time between a passion for science and the family.

The TERA Foundation and CNAO, the Italian National Project for Hadron Therapy

In 1992, the idea of an Italian hadron therapy centre began to take shape, but the planning of a centre for protons and light ions advanced slowly because of the few staff available. Tosi and I decided therefore to adopt the proposal of the brilliant 31 year old Gaudenzio Vanolo, who suggested to create a Foundation and later became the indefatigable Secretary General of the TERA Foundation (TERA being the Italian acronym for 'Therapy with Hadronic Radiation").

Over more than 20 years TERA has benefited from contributions of more than 200 young physicists and engineers; most of them are presently working in prestigious laboratories and hospitals. With the first collaborators we designed a hadron therapy centre – which we called from the beginning 'National Centre for Oncological Hadron Therapy' (CNAO) whose beating heart is a synchrotron for protons and carbon ions.

We had 10 years of ups and downs, of designs completed (first for the small town Novara and then for the city of Milan) and ultimately rejected, disappointments and strokes of luck. Finally in 2001 the CNAO Foundation was created by the Health Minister Umberto Veronesi, a world renowned oncologist, who granted initial finance of 20 million euros. Between 2002 and 2003 the young TERA staff worked on the definitive version of the project under the direction of Sandro Rossi (Fig. 8.11 on p. 233), who managed the construction of the centre and became CNAO Secretary General.

For the CNAO design we benefited from a study carried out at CERN between 1996 and 2000, the so-called PIMMS (Proton-Ion Medical Machine Study), which was led by the English engineer Phil Bryant, a very experienced accelerator scientist. TERA initiated it and contributed, together with the Austrian MedAustron group and many CERN scientists and engineers, who made their expertise available. The final publication was conceived as a toolkit from which any European user could select and adapt the parts best suited to its own goals.

In producing the construction drawings of CNAO we chose to keep the features of the PIMMS synchrotron but to reduce the footprint of the centre as much as possible. Thus, as shown in Fig. 8.15, a single injection linac and the proton and carbon ion sources were located inside the ring and short magnetic lines were designed to transport the beams from the synchrotron to the three treatment rooms. No gantry was built, but space was left for a later installation of two carbon ion gantries, as large as the one of HIT (Fig. 8.14).

The construction of CNAO started in 2005 on land adjacent to the San Matteo Hospital in Pavia, a university town 30 km south of Milan. It was made possible by the funds allocated to the CNAO Foundation by the Health Ministry and generous contributions from Italian institutions, in particular INFN, the National Institute for Nuclear Physics, which built many high technology components.

The first patient, a young man of 26 already operated on twice for a chondrosarcoma at the base of the skull, was irradiated in September 2011 and

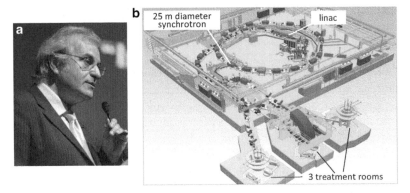

Fig. 8.15 (**a**) Roberto Orecchia is responsible of the Radiotherapy department of the European Oncological Institute (Milan) and CNAO Scientific director (Courtesy of CNAO Foundation – Pavia). (**b**) The CNAO synchrotron is 25 m in diameter and, with very short magnetic transport lines, serves three treatment rooms, the second of which features a horizontal and a vertical line (Courtesy of Istituto Europeo di Oncologia, Milan)

returns from time to time to visit the centre. After HIT, CNAO was the second institution in the world to offer – outside Japan – radiotherapy treatment with carbon ions.

In the framework of ENLIGHT – the European Network for Light Ion Therapy coordinated at CERN by the biophysicist Manjit Dosanjh – HIT and CNAO have joined forces to develop common protocols and compare the effects of proton and carbon ion treatments on a number of tumours. The general framework of this collaborative effort is a study, published in 2004 by ENLIGHT, in which Austrian, French, German and Italian radiation oncologists concluded that, of the 2,000 patients who are treated with X-rays every year in a population of one million inhabitants, about 12 % (240 patients) would profit from a proton treatment and about 3 % (60 patients) of a carbon irradiation.

In 2014, with HIT and CNAO running, and the centres in Marburg (Germany) and Wiener Neustadt (Austria) soon running, Europe is well placed to perform coordinated therapy comparisons on many patients, so as to compare the clinical effects of carbon ions with protons and define those tumours that are the optimal targets of this unique approach to radiotherapy. As explained, Japan remains at the forefront of this therapy, while in 2013 – 20 years after the closure of the Berkeley Bevalac – the National Cancer Institute and the Department of Energy in the USA have decided to launch a study for an American carbon ion and proton centre.

Present and Future of Radiotherapy

If the linear accelerators today used for medical applications in the entire world (about 20,000) were arranged one after the other, they would form a 30 km line, not much longer than the tunnel at CERN which today holds the Large Hadron Collider.

X-ray radiotherapy becomes ever more widespread and at the same time more accurate, and the number of linacs is increasing rapidly; the rate of growth is 2 % per year in the USA and 12 % in China, which already in 2011 had 1,500 linacs in use.

In future many more will be needed. In the western world on average 2,000 patients per million inhabitants are irradiated each year; one linac can treat 300 patients a year, and international standards would consider provision of about one linac for every 150,000 inhabitants to be sufficient. According to these criteria, China would need another 8,000 linacs.

Linear accelerators are employed in 90 % of radiation treatments and, on average, every patient is irradiated 30 times. Three types of cancer – lung, breast and prostate – account for about 60 % of all treatments, which are often preceded or followed by surgery or chemotherapy. The 5-year survival rate is 70 % in the case of phase-1 lung cancers, which are those tumours localised in the lungs which have not spread to nearby lymph nodes or metastasised in other organs.

Today the advantages of proton therapy are recognised by many radiation oncologists. The argument is that protons have the same biological and clinical effects as X-rays while concentrating the dose much better on a tumour target. However it has to be stressed that this is not the opinion of all radiation oncologists. Other radiotherapists argue that the large investments and higher costs per patient (by a factor 2–3) are not sufficiently justified by the accumulated clinical results, particularly in view of the lack of direct comparison between X-ray and proton outcomes in so-called 'phase III clinical trials'. The issue is hotly debated in scientific journals and, I am convinced, will be resolved only when many hundred thousand patients will have been treated.

The best way to utilize the *macroscopic* advantage of protons is to impart the same dose to the organs at risk as conventional X-ray irradiation but increase the dose to the tumour, thereby augmenting the control probability rate of the irradiated cancers. More generally protons may be used to increase tumour control, or reduce normal tissue complication – in particular the number of late radiation-induced tumours – or some combination of the two.

The increasing awareness of these simple facts is demonstrated by the exponential world growth of the number of proton therapy centres, which multiply by a factor 2 every 10 years. In the United States in 2011, ten centres were in operation and more than eight under construction; and in that year 600 children and adolescents were irradiated with protons, an increase of 30 % compared to 2010.

Europe, which is at the frontier in carbon ion centres, moves more slowly in the clinical application of protons, with seven centres in operation and three under construction, even though the 2004 study by ENLIGHT concluded that, of 2,000 patients treated with X-rays per million inhabitants, around 240 would benefit from treatment with protons.

The required investment for a proton therapy centre is greater than 100 million euros but – taking account both the expense, as well as the number of patients treated each year – in Europe the cost of a complete proton therapy course varies between 20,000 and 25,000 euros, to be compared with 8,000–12,000 euros for precision therapy with X-rays and with 50,000–70,000 euros for many extended chemotherapies. American costs are more than double this amount.

For fundamental physics three main types of accelerators have been invented – cyclotrons, synchrotrons and linacs – but only the first two are today used for light ion therapy. This is now changing because the TERA foundation has designed and prototyped a novel linac which can accelerate protons from 30 to 230 MeV in about 15 m. The CERN spin-off company A.D.A.M. (Applications of Detectors and Accelerators to Medicine) in Geneva is building the first centre based on the new technology, which allows active spreading of the dose with 'multiple-painting' of the tumour target.

It must be strongly underlined that proton therapy could almost completely replace X-ray therapy – in every hospital, for 90 % of solid tumours – if only it was possible to build 200–250 MeV proton accelerators as small (a couple of metres long) and 'cheap'(one to two million euros) as 10 MeV electron linacs.

This technological goal is ambitious, but specialised firms (like the Belgian IBA and the American MeVion) and research laboratories already offer cyclotrons compact enough to be mounted on a headpiece of 8–10 m in diameter, which rotates around the patient. The TERA foundation has also been working on the design and realization of a proton linac mounted on a rotating gantry. Moreover laser accelerators, potentially smaller than cyclotrons, synchrotrons or linacs, are also under study to reduce the dimensions of hadron therapy facilities; but more than 10 years will be needed to bring these laser research projects from the laboratory to the clinic.

Treatment with carbon ions produce *macroscopic* dose distributions similar – but with sharper edges – to those of protons. However, carbon ion *microscopic* doses are very different because ions are a densely ionizing radiation; this opens the way to the control of tumours that are radioresistant to both X-rays and protons. In the next 10 years many clinical trials will have to be performed in which the outcome of proton and carbon ion treatments will be compared and finally it will be possible to clearly identify the tumours for which carbon ions are proven to be more effective.

The Landscape

For 120 years the effects on medicine of advances in particle accelerators for fundamental physics and, vice versa, the positive stimuli of medicine on the development of new and more efficient machines have been continuous and profitable to both fields – medicine and physics.[9] The cases described in the last

---

[9] Good examples of the influence of medical applications are the 60 in. cyclotron built by Lawrence for producing medical isotopes and the developments of electron linear accelerators pushed by their use in cancer therapy.

Fig. 8.16 The three-yarn thread is used as timeline of the events listed in Table 8.2

two chapters are classic examples of positive impact from science which, starting from 'pure' research, produces benefits for the whole community. Cyclotrons, synchrotrons and linacs – the three types of accelerator invented to carry out experiments in nuclear and particle physics – are today instruments which help physicians to improve diagnoses and treatment of diseases, increasing the duration and enhancing the quality of life of the seriously ill.

Guided by the invention and development of these instruments, we have followed the 120 year long path which brought us from X-rays to the discovery of the Higgs field and to hadron therapy. To complete the picture, in the Epilogue the interested reader has the opportunity to go back to the photo of the cosmic relic radiation – shown in the Prologue – and see how the knowledge gathered with particle accelerators has been used to reconstruct what happened in the cosmos from one millionth of a millionth of a second onwards.

It is now time to close the main text, take up again the three-yarn thread and recall the milestones that punctuated our trip.

Figure 8.16 highlights the dates of the relevant turning points in the history of particle accelerators, which have been narrated in this book and which are separated by time intervals of about 15 years. The main events and their protagonists are listed in Table 8.2, which covers the 60 years up to 1954, when the first proton therapy treatment in Berkeley and the creation of CERN occurred.

The following 60 years have been at least as successful, but they cannot be fitted in a neat pattern of events separated by about 15 years and are not easily summarized in a table. For this reason there are no entries in the table after 1954 and in Fig. 8.16 only two dates appear: 2012–2014. Of course the first one recalls the discovery of the Higgs field, a major event belonging to the black yarn. The second date refers to the creation – from January 1, 2014 – of the '*Office for CERN Medical Applications*' directed by Steve Myers, who was responsible for the Large Hadron Collider during its operational years (Fig. 3.11 on p. 79). In collaboration with physicians and medical physicists of more than 20 European universities and hospitals, the Office will coordinate and promote research activities on detectors for medical imaging, on software applications for medicine and on accelerators for tumour therapy. This guarantees that also in future the 60-year-old laboratory will continue to contribute to both the red and blue yarns.

Table 8.2 The discoveries and inventions that have marked the first 60 years of particle accelerator history are listed together with the main protagonists

Years	Black yarn: fundamental physics	Red yarn: therapeutics	Blue yarn: diagnostics	The protagonists
1895–1898	X-rays radium	X-rays radium	X-rays radium	Röntgen the Curies
1912–1913	Cosmic rays	Fractionation	X-ray tube	Hess Coutard Coolidge
1929–1932	Cyclotron Positron	Cyclotron	Cyclotron PET	E. Lawrence J. Lawrence Anderson
1945–1947	Synchrotron Linacs	Synchrotron Linacs Protontherapy	Linacs	McMillan Veksler Hansen S. Varian R. Varian Wilson
1953–1954	Creation of CERN	Proton therapy		Rabi Auger Amaldi Tobias

Epilogue

CERN Retraces the Course of Time

In closing I return to Fig. 1 of the Prologue and to the map of cosmic background radiation constructed by the Planck probe, the 'photograph' of the oldest state of the universe which has ever been made and which ever can be made. It goes back to 380,000 years after that explosion of energy and space which is called the Big Bang, initially only as a joke, and which says that the primordial plasma which made up the universe was practically all at the same temperature.

The background radiation, which since then has pervaded space, moving with it in every direction, is therefore a 'relic' of a remote epoch; although it informs us of the state of the cosmos in that era, and tells us something about what happened beforehand, it hides the first part of the story from our view.

To learn about the beginnings of the universe we must go back to what we have understood, thanks also to particle accelerators, about interactions between the fundamental particles; in this way we can attempt to reconstruct with the most powerful computers what happened from the Big Bang up to 380,000 years later, the era in which the universe became transparent.

Retracing in time the process of the separation of the galaxies and studying the cosmic background radiation in detail, astrophysicists have reached the conclusion that the universe came into being 13.8 billion years ago.

The initial temperature, at the beginning so high as to be unpronounceable, fell – initially very rapidly and then more slowly – as the universe expanded, until it had reached 'only' 3,000° after 380,000 years, when the universe was made up only of the simplest elements: 92 % hydrogen (one proton and one electron) and 8 % helium (a nucleus made of two protons and two neutrons, with two electrons rotating around them).

As already noted, temperature is proportional to the average energy with which the particles of a body are excited and collide; to heat water in a saucepan, for

© Springer International Publishing Switzerland 2015
U. Amaldi, *Particle Accelerators: From Big Bang Physics to Hadron Therapy*,
DOI 10.1007/978-3-319-08870-9

Fig. 1 In this photo, obtained in 2013 by the ESA Planck space probe, the different colours represent tiny fluctuations in the temperature of the gas, mainly hydrogen, of which the universe was made 380,000 years after the Big Bang (Courtesy European Space Agency, Planck Collaboration)

example, means to increase the average energy with which the water molecules, moving around, bump into one another.

The concept of 'energy with which particles collide' is simple and intuitive, such that cosmologists, to measure temperature, prefer to use a unit of energy instead of degrees; for them the unit of temperature is 1 GeV, the same unit which serves to measure the energy and mass of particles. In these units, the temperature of 10,000° (which the gas of hydrogen and helium had just before the famous 380,000 years and which today is characteristic of the surface of many stars) is equal to a billionth of a GeV.

In the following I will indicate with T the 'universal' time passed since the Big Bang. Thermodynamic considerations regarding the primal plasma allow to write the formula which links the temperature of the primordial universe – that is the average energy in collisions between particles – to the time T when these collisions occurred.

If E is measured in GeV and T in microseconds (i.e. millionths of a second, 10^{-6} s), the relation takes an easily memorisable form: $T = 1/E^2$. This says that one microsecond after the Big Bang the average energy of collisions was 1 GeV, while after one second, i.e. 10^6 microseconds, the temperature had fallen to a thousandth of a GeV (since $1/(10^{-3})^2 = 10^6$).[1]

---

[1] This simple formula is valid if most of the energy of the universe resides in light particles, mainly photons and neutrinos, which move at the speed of light. This was the case until about 50,000 years after the Big Bang, during what is called the 'radiation dominated era'. Afterwards the energy density due to massive, slowly moving particles became larger than the energy density of photons and neutrinos and the universe entered the 'matter dominated era'. For simplicity, in what follows I apply the same formula up to 380,000 years.

Going Back in Time with the Cosmic Hyper-oven

The composition of the primordial 'cosmic soup' changed with the passage of time. To explain why and how, let us imagine operating a 'hyper-oven' so well isolated, and made of material so heat resistant, that its temperature can be increased without limit.

Our hyper-oven contains the same mixture of hydrogen and helium as the recently transparent universe. Figure 2 shows what happens when we go *backwards in time*, heating the hyper-oven, starting from the 380,000 year old cosmos at 3,000°.[2]

In this back-to-front history, with the increase of temperature collisions between atoms become so violent as to detach all the electrons from nuclei; at a temperature of 10^{-7} GeV (one million degrees) the neutral gas has thus become an *atomic plasma* made of freely moving charged particles, and therefore opaque to light.

Continuing to go back in time, when the temperature increases to 0.01 GeV, the composition of the plasma remains unchanged; all that happens is that protons, electrons and helium nuclei collide more and more violently.

However, above 0.01 GeV (100 billion degrees), the collisions become violent enough to overcome the nuclear force which binds protons and neutrons together in

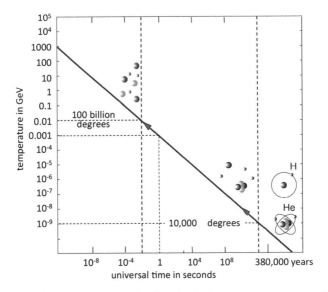

Fig. 2 By heating a gas of hydrogen and helium in the hyper-oven to a continually increasing temperature (i.e. ascending along the vertical axis) it is possible to reproduce the physical phenomena which took place at ever more remote times (i.e. moving to the left on the horizontal axis)

---

[2] Since the density of the universe was decreasing while the universe was expanding, in going back in time the cosmic soup should be heated and also compressed. The hyper-oven is a model of an expanding cosmos which is simplified, but adequate for the explanation.

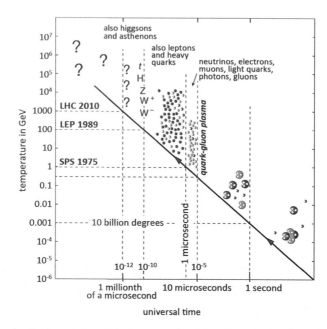

Fig. 3 Increasing the temperature of the hyper-oven, and therefore retracing the course of time, the composition of the cosmic soup changes. First, around 0.3 GeV, protons, neutrons and their antiparticles melt forming a quark-gluon plasma. At higher temperatures matter-particles and force-particles (with their antiparticles, not mentioned for space reasons) of ever-increasing mass appear

helium nuclei. As can be seen in Fig. 2, after exceeding 100 billion degrees the plasma changes composition; now it is made only of protons, neutrons and electrons, with no more helium nuclei.

When the temperature of the hyper-oven increases 30 times and approaches 0.3 GeV the protons and the neutrons (i.e. the nucleons) of the cosmic soup strike one another with such energy that a new phenomenon reveals itself; the collisions disassociate the nucleons liberating their components: the gluons and light quarks, i.e. u-quarks and d-quarks.

This special state of matter is called the '*quark-gluon plasma*' since in physics a 'plasma' is a state in which the constituent particles are electrically charged. According to the relation $T = 1/E^2$, this transition happened in the primordial universe when the universal time equalled 10 microseconds, i.e. 10^{-5} s. At this time the universe was about as big as our solar system and the typical distance between the nucleons was less than 1 fm, which corresponds to the diameter of a nucleon, and the cosmic density was greater than the already enormous one of a nucleus, 10^{15} g per cubic centimetre.

This change of composition is shown in Fig. 3, which is an enlargement of the left part of Fig. 2. The lightest circular symbols shown close to the name 'quark-gluon plasma' represent light quarks, electrons, neutrinos, gluons and photons. The appearance in the hyper-oven of this new state of matter is the *first* manifestation of

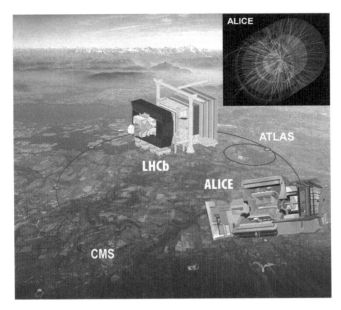

Fig. 4 ALICE is the LHC detector which is optimised for the study of phenomena which occur in collisions between two nuclei, which cause the creation of a plasma of quarks and gluons. The insert shows an event resulting from a collision between two lead ions (Courtesy CERN)

the many completely new phenomena which occur when matter is brought to higher and higher temperatures.

The quark-gluon plasma contained in the hyper-oven is made up only of *light* matter-particles and antiparticles (neutrinos, electrons, u-quarks and d-quarks with their anti-particles) which all move and collide with the same average energy of 0.3 GeV.[3]

At CERN the quark-gluon plasma is studied by an international collaboration of about 1,200 physicists originating in 36 different countries, which designed and constructed ALICE, a particle detector specialised for this purpose. ALICE is, after ATLAS and CMS, the third LHC detector in size and located at one of the interaction points of the 27 km ring (Fig. 4).

For several weeks each year in the Large Hadron Collider lead ions are collided against other lead ions, instead of protons with protons, precisely to study this unique state of matter. When in 2010 two 2,800 GeV lead ions – counter-rotating in the vacuum chambers of Fig. 6.3 (p. 156) – collided head-on, the 5,600 GeV was shared among about 600 quarks in a volume which had a diameter of about 8 fm.[4]

---

[3] At this temperature, the fraction of s-quarks in the plasma is small, both because they decay in less than a millionth of a second into light quarks but also because the energy available is insufficient to create many of them.

[4] The counting goes as follows. Each lead nucleus accelerated by the LHC is made of 208 nucleons (82 protons and 126 neutrons), i.e. of $3 \times 208 = 624$ quarks. Since, as depicted in Fig. 6.4 on p. 157, half of the energy of the two colliding nucleons (5,600 GeV) is carried by gluons, which keep the quarks together, each quark carries on average about $2,800/624 = 4.5$ GeV.

In the collision hundreds of new light quarks and antiquarks were created and a high-temperature quark-gluon plasma was formed. This bubble of plasma cooled down rapidly as all the particles and antiparticles forming it migrated away from the collision point at almost the speed of light (inset of Fig. 4).

With its initial data the ALICE detector confirmed what had been discovered in 2005 in gold-gold ion collisions at Brookhaven National Laboratory in the United States; the quark-gluon plasma, which before was thought of as a very dense gas, behaves like a liquid without viscosity. The large amount of data collected in subsequent years has been used to understand the detailed behaviour of the quark-gluon plasma and to check that its properties are in agreement with the predictions of quantum chromodynamics, the theory at the heart of the Standard Model of quarks and gluons.

From a Microsecond to a Millionth of a Microsecond, and Vice Versa

When the temperature in the hyper-oven increases above 1 GeV, the particles of the quark-gluon plasma collide with continually increasing violence; the energy liberated in each collision is now sufficient to produce *new* particle-antiparticle pairs, heavier than u-quarks, d-quarks, electrons or muons. So tau leptons and heavy s, c, b and t quarks appear in the cosmic soup (represented by the dark circular symbols of Fig. 3). Among matter-particles, those which appear at the highest temperature are top and antitop quarks, which have masses equal to 175 GeV.

Among force-particles at higher and higher temperatures, the W and Z asthenons, which have masses of the order of 100 GeV, are added to gluons and photons and Higgs bosons, with a mass of 125 GeV.

When the hyper-oven exceeds a temperature of 300 GeV (corresponding in universal time to 10^{-11} s) the Higgs condensate evaporates, like water at the bottom of a container raised above 100°. The full-vacuum of the hyper-oven changes state and *all* the types of Standard Model particle and antiparticle move within it, without any longer interacting. In other words, the 37 fundamental quantum fields are now all excited and the 24 types of matter-wavicles and 12 types of force-wavicles are present, but however without mass and moving at the speed of light.[5]

In the constant collisions new particles and antiparticles are created, but – at a given temperature – the number of wavicles of each type does not change, because a condition of *dynamic equilibrium* is established between the particles and antiparticles which are created and those which decay or, colliding randomly, are annihilated.

What happens far above 1,000 GeV? The scenario is still unknown from the experimental viewpoint, as the question marks of Fig. 3 show, but theoretical models like SUSY predict the existence of numerous new, still undetected particles (superparticles? Other types of exotic particle?).

---

[5] For a reason that cannot be simply explained, even without Higgs condensate the higgsons still have a mass.

Moving *from right to left* in Fig. 3 one can understand how the composition of the cosmic soup changes when the temperature is forced to increase. The primordial universe underwent the sequence of events described in figure *from left to right*, because while expanding it cooled down.

Before 10^{-12} s after the Big Bang, when the energies exchanged exceeded 1,000 GeV and the Higgs condensate did not exist, the asthenons had no mass, like photons; photons and asthenons were emitted and absorbed in the same way and the electromagnetic and weak forces were different aspects of the same unified force, the *electro-weak force.*

With the reduction in temperature the Higgs condensate appeared, at around 10^{-11} s, and gave mass to all particles and antiparticles, breaking the symmetry between the electromagnetic force and the weak force. This phenomenon is called 'electro-weak symmetry breaking'.

With the passage of time heavy force-particles, including the Higgs boson, and then leptons and heavy quarks (with their antiparticles) disappeared by annihilating and decaying – in a fraction of a microsecond – into lighter particles.

These heavy unstable particles did not remain in the matter that is around us; atoms are thus made only of *u*-quarks, *d*-quarks and electrons, which are too light to decay into something else.

To discover the existence and study the properties of these missing particles it was necessary to accelerate infinitesimal fractions of matter and antimatter to ever-higher energies; colliders are, in good part, a substitute for our imaginary cosmic hyper-oven.

The text SPS, LEP, LHC on Fig. 3 shows how in the last 40 years the CERN accelerators have provided – and still provide today, with the LHC – information on the composition of the cosmic plasma at times closer and closer to the Big Bang. This is why it can be said that, with its ever more powerful colliders, *CERN retraces the course of time.*

The precision measurements obtained with particle accelerators permit identification of all the complicated processes which happen when either one electron annihilates with a positron (as in LEP) or a proton collides with a proton (as in the Tevatron and, at larger energies, in the LHC) – thus allowing the construction of the Standard Model which collects all the results in a coherent framework. Then, by using all the knowledge codified in the Standard Model, physicists *simulate* with powerful computers all the phenomena which happen in the cosmic soup first at 1,000 GeV and then at all lower temperatures, thus following the evolution of the universe in detail, from a millionth of a microsecond (10^{-12} s) up to 380,000 years after the Big Bang.[6]

---

[6] When one says that two colliding protons at LHC give us information on what happened 10^{-12} s after the Big Bang, this is true in a limited sense. ATLAS and CMS detect the particles and antiparticles created in the collisions between two protons, i.e. mainly between light quarks and gluons, collisions which were taking place 10^{-12} s after the Big Bang, but cannot directly reproduce, for instance, the annihilation between a top quark and its antiquark or the collision between a neutrino and a W-asthenon. However this is not necessary because the study of the observed collisions and decays of all matter-particles is integrated in the Standard Model, which can then be used to simulate on a computer all types of collisions and decays of all known particles and antiparticles, as they should have occurred in the primordial soup.

The good news is that in this way the image of Fig. 1 can be explained in its composition (95% hydrogen and 5% helium) and temperature (3,000°); moreover the very small temperature fluctuations are accurately reproduced. However, before reaching this very satisfactory result, two serious difficulties have to be overcome: the *horizon problem* and *the matter–antimatter asymmetry problem*. I discuss them in the next two sections.

The Primordial Universe and Inflation

To understand the 'horizon problem', it is useful to return to the image of background radiation of Fig. 1; we find ourselves at the centre of an enormous sphere, whose light reaches us after 13.8 billion years.

Looking at the spherical surface we see how the neutral gas, made of hydrogen and helium atoms, was distributed, when it was formed at 3,000° 380,000 years after the Big Bang. Different regions have slightly different temperatures, as shown by the colours in the figure; for example the blue spots were very slightly colder, by just 0.05°, compared to the 3,000° average.

In colder regions the density was slightly larger than average, but this miniscule difference was enough to ensure that, as a result of gravitational attraction, the cold points would become centres of hydrogen and helium concentration, so forming the earliest 'seeds' of stars and galaxies.

The almost perfect uniformity of the primordial gas is surprising, because it means that regions of gas in positions diametrically opposite on the enormous sphere must in some way have 'communicated with each another' before 380,000 years; how could they otherwise have the same temperature?

To understand where the difficulty lies, one can argue in the following way. When the photons of the background radiation were emitted, the universe was a sphere a thousand times smaller than today. In that epoch any signal which moved at the maximum possible speed, the speed of light, would have required 30 million years to travel the diameter of the sphere and allow communication between the two opposite regions of the universe, bringing them to thermal equilibrium. But since only 380,000 years had elapsed since the Big Bang, the photons had not had enough time to cover those 30 million light-years. The same reasoning applies for all earlier times; the cosmos has never been fully within the range of photons.

However, if we assume that at a certain moment there was a *very rapid and accelerated expansion* of the universe, then before the expansion the cosmos could have been small enough to allow traversal by 'equalising signals', which brought it to a state of thermal equilibrium.

Alan Guth, an American theoretical physicist, with his 'inflationary model' of the universe proposed the hypothesis of accelerated expansion in 1979. According to Guth's model, between 10^{-36} and 10^{-32} s the cosmos underwent an enormous and extraordinarily rapid increase in size, doubling in diameter approximately every 10^{-34} s, as a result of a sudden and powerful push from a scalar field, similar to the Higgs field, to which the horrible name 'inflaton' has been given.

During this *cosmic inflation* the diameter of the universe increased at least 10^{30} times, passing from a miniscule fraction of the diameter of a proton to about a centimetre. At the end of this process the energy of the inflaton had given rise to all the possible particles and antiparticles predicted by SUSY (if this theory is true) raising them to the enormous temperature of 10^{13} GeV.

Before inflation, inhomogeneities caused by fluctuations typical of quantum physics were present in the cosmic soup. During inflation these small variations in density and temperature were smoothed out, like the wrinkles in an empty balloon which suddenly swells up. This is why, after 380,000 years, the plasma had the same composition and density everywhere, perturbed only by those tiny fluctuations which became seeds of galaxies.

The Problem of Matter–Antimatter Asymmetry

The second serious difficulty in the reconstruction of the history of the early universe originates from this question: why, if in all collisions which occur in the primordial soup, a particle is always produced with an antiparticle, is the present universe made only of matter-particles, without any trace of matter-antiparticles?

It is natural to assume that, at the end of inflation, the electric charge of the still tiny universe was globally zero, which means that the number of positively charged particles was exactly equal to the number of those negatively charged.

The neutrality of the cosmos is also maintained after that time because in all phenomena that we know of, electric charge is conserved. A positive particle annihilates with its negative antiparticle producing a flash of energy, such that the total charge remains zero. And conservation of electric charge also holds in the process of particle decays, in which the sum of the final charges is equal to the charge of the decaying particle, and in every particle-particle collision.

Therefore it is expected that after 10^{-4} s from the Big Bang – at the end of the era dominated by the quark-gluon plasma – the number of (positive) protons and (negative) electrons should have been identical to the number of (negative) antiprotons and (positive) positrons; and that the number of neutrons was equal to the number of antineutrons. At those times temperatures were less than 0.1 GeV (Fig. 3) and the energy available in collisions was not sufficient to create new proton-antiproton or neutron-antineutron pairs, so that all the nucleons and antinucleons should have disappeared in a myriad of annihilations. Similarly, electrons and positrons, which could have been created while the temperature remained above 0.001 GeV, would have met the same fate around one second after the Big Bang.

In a few words, the cosmos ought to have had – throughout its history – a very natural form of symmetry: the 'matter–antimatter symmetry'. But, if it had been like this, *all* the particles and antiparticles should have annihilated leaving a universe made only of photons; a period should have existed which could be legitimately called the 'great annihilation'; the photons produced would then have continued to travel at the speed of light in an expanding universe completely *empty* of matter-particles.

Here is the difficulty; this did not happen because, 14 billion years later, the universe is the ensemble of a hundred billion galaxies, made up of around 10^{80} protons and neutrons (what we call 'matter') with an equal number of electrons, but without any trace of antiprotons and antineutrons (antimatter). How did it happen that the antimatter disappeared leaving behind a universe made only of matter?

Fortunately for us, at the time of the great annihilation in the cosmic soup there must have been a tiny prevalence of matter-particles compared to antimatter particles. The explanation of this asymmetry – now accepted by all physicists – was proposed in 1967 by the Russian physicist Andrei Sakharov who, in the 1980s, became a famous dissident within the Soviet Union regime. To explain it I will make use of a greatly simplified numerical example.

At the time of the great annihilation, 1,000,000.001 protons were found in the cosmic plasma for every 1,000,000.000 antiprotons, so that just a single proton survived while a billion protons annihilated with a billion antiprotons; the same thing happened to the 1,000,000.001 electrons initially present with the 1,000,000.000 positrons, so that finally a single atom of hydrogen remained, made of a proton and an electron.[7] In other words, at that time the matter–antimatter symmetry was not exact but 'broken', if only by a tiny amount.

How can we claim with such precision that the excess of protons and electrons was exactly one part in a billion? This number has been obtained by observing how many photons are today present in the cosmic background radiation (photographed in Fig. 1) compared to the number of protons. The photons are in fact the same ones that were produced in the era of particle-antiparticle annihilations. The energy of these photons has diminished enormously because they have travelled for 14 billion years in a perpetually expanding cosmos and the creation of new space has stretched out the crests of the electromagnetic waves, with a corresponding reduction of their energy. However, their number has remained essentially invariant. Today on average one photon of cosmic background radiation is found in every cubic millimetre of space while – always *on average* – there is one proton for every cubic metre of the universe. The value of one proton for every billion photons comes from the comparison of these two volumes.

In the model proposed by Sakharov, the charge asymmetry is then due to the fact that, in the very first instants after the Big Bang, some particles existed which, in decaying, produced more particles than antiparticles. A phenomenon of this kind was observed in 1964 at the Brookhaven *Alternating Gradient Synchrotron* where the decays of neutral kaons (made of a d-quark and an $\bar{s}$-antiquark) were compared with those of neutral antikaons (composed of an s-quark and an $\bar{d}$-antiquark). In recent years the LHCb detector of Fig. 4 has discovered a few decays of b-quarks and $\bar{b}$-antiquarks which also do not respect the matter–antimatter symmetry.

---

[7] Because the total electric charge should always be zero, the number of (negative) electrons is always equal to the number of (positive) protons and the number of (positive) positrons must be identical to the number of (negative) antiprotons.

The phenomena observed up to now in the laboratory have shown that there are decays in nature which do not respect matter–antimatter symmetry – and therefore the principle of electric charge symmetry – but they are not sufficient to explain the asymmetry of one part in a billion which is needed to explain a universe composed only of matter. Today, despite 50 years of experimental searches and theoretical speculation, we do not know which particles, with their asymmetric decays, have caused it; the problem of the matter–antimatter symmetry still awaits a solution.

However, many theoretical physicists believe that the asymmetry is due to the decay of particles that are *not* included in the Standard Model. In particular they refer to very heavy neutrinos, which should exist in many, logically coherent, extensions of the Standard Model.

The Four Forces Distinguished Themselves in the Course of Time

Today most physicists accept the mechanism proposed by Guth to resolve the horizon problem and the data collected by the Plank probe, concerning the cosmic temperature distribution of Fig. 1, confirm it.

Starting from this – and by assuming that the matter–antimatter problem is solved by the asymmetric decays of yet unknown particles – Fig. 5 gives a synthesis of what happened in the first 380,000 years. Both universal time and the temperature are plotted on the horizontal axis while the vertical axis shows the size of the universe.

The time scale begins with the *Planck time*, which is equal to about 10^{-43} s and represents the instant before which current fundamental physical theories, that is general relativity and quantum mechanics, are no longer valid.

The upper part of the figure, as well as showing the phenomena already illustrated in Fig. 3, demonstrates that the cosmic scene was dominated by *two* processes: *inflation* and *expansion*.

The lower part of the figure shows the state of the four fundamental forces. To understand the significance of this, it is necessary to recall that their couplings depend on the energy exchanged in collisions.

Figure 6.17 on p. 175, which illustrates the unification of the forces under the assumption that SUSY is valid, can be converted – using the relation $T = 1/E^2$ – to Fig. 6, where the lower horizontal axis represents universal time and the upper axis the temperature expressed in GeV.

The evolution of the couplings of the three forces as a function of universal time gives a more accurate picture, compared to the lower part of Fig. 5, of what happened between the Planck time and a millionth of a second after the Big Bang.

At the time of inflation the couplings of the electromagnetic, weak and strong forces had similar values and the exchange of gluons, photons, asthenons and gravitons occurred with the same probability; *the four forces were unified*, different

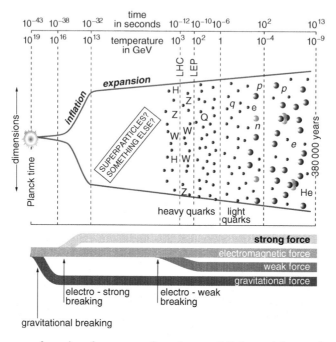

Fig. 5 By means of a series of processes of creation, annihilation and decays of particles and superparticles – if they exist – in 1 microsecond (10^{-6} s) after the Big Bang the temperature fell from 10^{19} GeV to 1 GeV; at that time only light quarks and electrons collided and were created in the soup. (For simplicity we have not shown either antiparticles or light neutral particles: gluons, photons and neutrinos)

manifestations of a single force whose force-particles were closely related (or, as physicists say, 'belonged to the same multiplet').

Even without a convincing theory of the unification of the electro-strong force with the gravitational force, it is reasonable to suppose that around 10^{19} GeV the unified coupling strength of the three forces, which act in the subatomic world, would have been equal to the gravitational coupling strength, which we know increases very rapidly with energy. For this reason, in the vicinity of the Planck time, Fig. 5 shows the hypothetical *gravitational breaking* between gravitation and the unified strong-weak-electromagnetic force.

Between 10^{-38} and 10^{-35} s, at temperatures of about 10^{15} GeV, the interval indicated in Fig. 6 by the grey band on the left, the strong and the electro-weak couplings began to diverge and the *breaking of the electromagnetic, weak and strong symmetry* – i.e. the electro-strong breaking – occurred. In the same era, the universe experienced the phenomenon of inflation, which drove it to a diameter of about a centimetre – after 10^{-32} s – and to begin a much slower expansion, as shown in Fig. 5.

If Supersymmetry is realized in nature – a fact that, as discussed at the end of Chap. 6, up to 2014 is *not* corroborated by LHC data – after 10^{-32} s *all* possible particles and superparticles were present (with their antiparticles), in a turmoil of

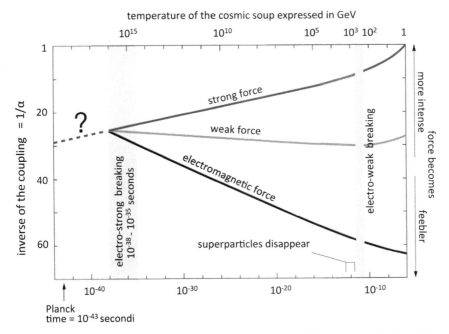

Fig. 6 The inverse of the couplings is plotted versus the universal time between the Planck time and one microsecond (10^{-6} seconds); the strong force becomes steadily more powerful, the electromagnetic more feeble. All the lines are calculated on the assumption that Supersymmetry is valid. The average energy E exchanged in collisions is identified with the collision energy E, i.e. with the temperature of the cosmic plasma

annihilations and creations at collision energies much larger than the masses of all the particles and corresponding superparticles.

10^{-11} s Is a Very Special Universal Time

The cosmos continued to expand until universal time passed 10^{-12} s and the temperature fell below 1,000 GeV. At that point a transition occurred which determined all the successive history of the universe; superparticles (and their antiparticles) – *if* they existed – stopped being created, because collisions with sufficient energy no longer took place.

Since supersymmetric particles are unstable, they decay rapidly into particles and antiparticles and into the lightest of all superparticles, generically called 'neutralinos', which – according to some theoretical models – can not themselves decay simply because there is no lighter superparticle which would make this possible. If these models are correct, eventually (so to speak, since this refers to millionths of

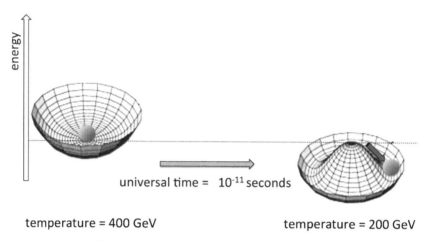

temperature = 400 GeV   temperature = 200 GeV

Fig. 7 At around 10^{-11} seconds the decreasing temperature induced a drastic change in the energy of the Higgs field which gave rise to the Higgs condensate filling the vacuum. Often physicists call the figure to the right a "Mexican hat"

microseconds) there only remained those stable neutral superparticles which did not bind to ordinary matter and therefore have since wandered throughout cosmic space.

As discussed in the next section, these stray neutralinos had, possibly, an essential function in the formation of galaxies.

Returning to the chronology of the primordial universe represented in Fig. 6, around 10^{-11} s the 'vacuum' changed, following the appearance of the Higgs condensate; since then the initial electro-weak symmetry has been hidden from the eyes of physicists and the electromagnetic and weak forces have very different properties.

Before this time, the mediators of the electric force (the photons) and of the weak force (the asthenons) were all massless and had almost equal couplings, so that they were equal members of the same family and their exchanges gave rise to different facets of a single unified force, the *electro-weak force*. How this can happen has been well described by Steven Weinberg.

The W and Z particles are not at all alike; the photon has no mass and the others are much heavier than any known particle. How can they be identified as members of the same family? The explanation for this is found in an idea originated in an entirely different discipline, the physics of the solid state. This is the idea of broken symmetry. Even if a theory postulates a high degree of symmetry, it is not necessary for the states described by the theory, that is, the states of the particles, to exhibit the symmetry. That sounds paradoxical, but let me give an example. A round cup is symmetrical around a central axis. If a ball is put into the cup it rolls around and comes to rest on the axis of symmetry. If it came to rest anywhere else it would mean that the solution of the problem violated the symmetry of the problem.

But one could have another cup in which a symmetrical dimple was pushed up into the bottom; in this cup a ball would come to rest at some indeterminate point in the round valley at the bottom of the cup, which breaks the symmetry because it is not on the axis of symmetry. Hence symmetrical problems can have asymmetrical solutions. This kind of broken symmetry is analogous to that evident in modern gauge theories. A better phrase would be hidden symmetry because the symmetry is actually present and one can use it to make predictions, including the very existence of the weak force. In this particular example, one can use the symmetry to predict the way the ball will oscillate if it is perturbed; in unified gauge theories of the weak and electromagnetic interactions, one predicts the existence of the interactions and many of their properties. Nothing in physics seems so hopeful to me as the idea that it is possible for a theory to have a high degree of symmetry which is hidden from us in ordinary life. (Weinberg 1977)

In Fig. 7 this analogy is applied to the appearance of the Higgs condensate in the universe which, expanding and cooling, passes from a temperature above 300 GeV to lower values; the ball (which represents the state of the Higgs field) moves from a symmetric state, represented by the base of the bowl on the left, to an asymmetric position because the lowering of temperature 'pushes up a dimple into the bottom', as Weinberg explains.

As shown in Fig. 7, it would be more correct to say that "the decreasing temperature pushes *down* the cup so that a symmetric dimple is left behind" rather than "the temperature pushes *up* a dimple into the bottom." In this way the ball keeps initially the same energy but, being in an unstable equilibrium, rolls *naturally* from the initially symmetric position to an asymmetric position on the symmetric bottom of the modified cup.

In rolling down, the ball chooses only one of the infinite final positions that are available to it in the circular groove of the Mexican hat. This choice breaks the symmetry the system had when the ball was sitting on the top of the dimple. Ed Witten – the main actor on the stage of string theories – has compared this phenomenon to the freezing of a liquid.

If weak interactions are so similar to electromagnetism, why do they appear so different in everyday life? According to the Standard Model, the key is 'symmetry breaking'. Even if the laws of nature have a symmetry – in this case the symmetry between the photon and W and Z bosons – the solutions of the equation may lack that symmetry.

For example, in a liquid, an atom is equally likely to move in any direction in space – there are no preferred coordinate axes. But if we cool the liquid until it freezes, a crystal will form. Which has distinguished axes. All directions in space are equally possible as crystal axes, but when the liquid freezes, some distinguished axes will always emerge. The symmetry between the different directions in space has been lost or 'spontaneously broken'. Similarly, according to the Standard Model, just after the Big Bang there was a perfect symmetry between the photon and the W and Z bosons. At the high temperature that then existed, electromagnetism and the weak forces were equally obvious. But as the universe cooled, it underwent a phase transition, somewhat analogous to the freezing of a liquid, in which the symmetry was 'spontaneously broken'. The W and Z bosons gained mass, limiting the weak forces to nuclear distances and putting their effect out of reach of the unaided eye. The photon remained massless, as a result of which electromagnetic effects can propagate over human-scale distances (and beyond) and are obvious in everyday life.

At high temperature, Higgs particles, like other particles, moved at random. But as the Universe cooled, Higgs particles combined into a 'Higgs condensate', an ordered state in which many particles share the same quantum wave function. The electroweak symmetry was broken by the 'direction' of the Higgs condensate (in an abstract space that describes the different particle forces) in roughly the same way that in a crystal, the rotational symmetry is broken by the direction of the crystal axis. (Witten 2004)

Thus the electromagnetic and weak forces began to be different when the temperature fell below 300 GeV and the Higgs condensate gave masses to the asthenons and brought about the 'breaking of electro-weak symmetry' shown in Fig. 6 by the grey band on the right.

When the universal time passed 10^{-10} s, the asthenons, the Higgs particles, top quarks (and antiquarks) disappeared because, once they had decayed into lighter particles, they could no longer be recreated in collisions, because too little energy was available. After 10^{-10} s the cosmic plasma was no longer composed of two dozen particles and two dozen antiparticles, but only a dozen matter-particles accompanied by their antiparticles; the e, μ and τ leptons, light and heavy quarks (except the top quark) and neutrinos, all with their respective antiparticles. The gluon and photon force-particles were also present, but for simplicity do not appear in Fig. 5 (where the heavy quarks are indicated by "Q").

The dynamic equilibrium between these particles did not last long because, in fractions of a millionth of a second, the heavy particles decayed into light particles and then, around 10^{-5} s, the plasma of quarks and gluons was formed. As we have already seen, in the subsequent second – after the 'great annihilation' – the light quarks gave way to protons and neutrons, which then bound with protons to form the first helium nuclei.

Dark Matter and Dark Energy

Through the hypothetical neutralinos Supersymmetry may have an important role not only at the beginning of the universe but also in the formation of galaxies.

The reason is that the observations of galaxies over a wide range of distances has, in fact, led astrophysicists to conclude that what is called 'visible' matter, that is the stars, interstellar dust and light – is not present in sufficient quantities to explain how peripheral stars rotate around the centres of their galaxies; if there were not something else present, the stars far from the centre should travel with rotational speeds many times lower than what is observed.

To explain these incontrovertible experimental measurements, it is generally accepted that, as well as visible matter, the volume of every galaxy is occupied by a new type of (electrically neutral) dispersed and invisible form of matter which must make up 85 % of the total mass of the galaxy. The visible mass is therefore only 15 % of the total.

What is the cloud of 'dark matter' made of? Many of the proposed solutions have been dismissed; the main idea, which still survives today, is that it is constituted by neutral particles which interact very little with the rest of matter. The most popular candidates are neutralinos, which would have been created in the initial moments of the universe and, in some theories, do not decay into other particles and superparticles; wandering through space, they would still be gravitationally bound to visible matter and contributed to the formation of galaxies.

The discovery of supersymmetric particles could not only enlighten us about the nature of dark matter, through neutralinos, but also about another surprising feature of our universe. From the picture of the relic radiation taken by the Planck probe

(Fig. 1 on p. 248) it can be computed that dark matter and visible matter represent *only* 32 % of the energy of the universe. What is the origin of the remaining 68 %?

In 1998 two groups of American astrophysicists found, that in the last five billion years the expansion of the universe has been accelerating. This very unexpected discovery has disconcerted the scientific world because it is natural to think that the gravitational attraction among the masses of all the galaxies should have been decelerating the expansion, pulling the universe back inward, just as gravity pulls a stone back down to Earth after it has been thrown into the air.

The best explanation today is that a 'dark energy' exists, which has a repulsive effect and is distributed *uniformly* throughout space, and for the last five billion years has forced the galaxies to move further apart from one another with increasing speed, pushing further the expansion initiated with the Big Bang. In other words, when the universal time became larger than 10 billion years, the expansion rate of the universe changed because the dark energy density became larger than the energy density of the massive slowly moving particles; this time marked the end of the 'mass dominated era' and the beginning of the 'dark energy dominated era'.

In summary, today the largest fraction of the universe energy is in the form of repulsive dark energy – which explains 68 % of the total. The remaining 27 % is dark matter (since 85 % × 32 % = 27 %) and *only* 5 % (i.e. 15 % × 32 %) is visible energy (or matter).[8]

Universe or Multiverse?

The origin of dark energy is, after the nature of dark matter, the second fundamental astrophysical problem waiting for a convincing explanation. String theory also has a lot to say on the second open problem.

Dark energy is the *energy of the cosmic vacuum*, or what remains in a volume of intergalactic space when the few particles which are found there are removed – along with all the electromagnetic waves and neutrinos which traverse it.

According to quantum physics, the *cosmic vacuum* is not simply *nothing* because it is 'filled' by the Higgs condensate, which breaks the symmetry between the electromagnetic and weak forces, and the continuously varying fluctuations of the 24 fundamental fields.[9] The energy density of this condensate can be estimated quite simply, with the result that it weighs very much more than the experimentally measured dark energy; the

---

[8] Today the measured *density* of dark energy is about 3 GeV per cubic metre. Since it is supposed not to vary with time, its contribution to the total energy of the universe was unimportant 380,000 years after the Big Bang, when the radius was 1,000 times smaller than now. Then dark matter dominated – contributing 63 % of the total energy – with almost equal amounts of photons (15 %), matter (12 %) and neutrinos (10 %).

[9] A quantum field which does not contain energy fluctuates randomly and continuously because of the uncertainty principle. In the language of particles this is expressed by saying that virtual particle-antiparticle pairs continuously appear and disappear from the 'vacuum'.

related gravitational repulsion is so large that the size of the universe should double every 10^{-38} s! The contribution to the energy density from the fluctuations of the other Standard Model fields further reduces this already incredibly short time by a big factor.

The problem is apparently *insoluble*; the theories constructed by particle physicists explain with marvellous precision the thousands of subatomic reactions observed at the Large Hadron Collider but, when they apply the calculations to dark energy they give a value which is too large by a factor of *many, many* orders of magnitude compared to what has been experimentally determined from the accelerated expansion of the universe, which began to dominate five billion years ago.

How can string theory resolve this contradiction? The remedy is actually in the number of possible realisations of the theory, which are *measurelessly* large. In each case, dark energy has a different value, sometimes large and, much less frequently, very small, as in the case of our universe. The existence of the cosmos as we know it is therefore an extremely fortunate event. To explain this extraordinary fortuitousness, in 1987 the Nobel Laureate Steven Weinberg appealed to the 'anthropic principle', which maintains that – as living beings – we cannot expect that our universe could have physical properties in contradiction with the development of life. Weinberg therefore argued that we measure such a small value of dark energy because, if it were larger, the universe would have rapidly collapsed and there would not have been the time necessary for the formation of stars and planets, like the Sun and the Earth.

More specifically, the anthropic principle maintains that our universe – in form, dimensions, age, fundamental structures which we observe, and phenomena which take place – must be such as to permit the evolution of conscious observers because, if intelligent life were not to emerge, there would be no means to pose questions on the form, dimensions, age, structures and phenomena of the cosmos. Mankind therefore has no reason to be surprised by the fact that structures and phenomena *are not observed* in nature which would *prevent* the appearance and evolution of life.

Yet, from here, it is a small step to the hypothesis that there should be an enormous number of different universes, which correspond to all possible manifestations of string theory (i.e. of different 'vacua') and in which the density of dark energy assumes every possible value. Sentient beings can then emerge only among the cold bodies of those *very few* universes that, like our own, are characterised by *unusually small* value of that quantity (Fig. 8).

In 2005, in a widely cited conference entitled "Living in the Multiverse" Steven Weinberg magisterially summarised the change of perspective.

> *Most advances in the history of science have been marked by discoveries about nature, but at certain turning points we have made discoveries about science itself. These discoveries lead to changes in how we score our work, in what we consider to be an acceptable theory. For an example just look back to a discovery made just 100 years ago.*
>
> *In the Special Theory of Relativity Einstein offered a symmetry principle, which stated that not just the speed of light but all the laws of nature are unaffected by the transformation to a frame of reference in uniform motion. Never before had a symmetry principle been taken as a legitimate hypothesis on which to base a physical theory.*
>
> *The development of the Standard Model did not involve any changes in our conception of what was acceptable as a basis for physical theory. Similarly, when the effort to extend the Standard Model to include gravity led to widespread interest in string theory, we*

Fig. 8 An artist's view of
the Multiverse (Mehau
Kulyk, Science Photo
Library)

*expected to score the success or failure of this theory in the same way as for the Standard
Model: String theory would be a success if its symmetry principles and consistency
conditions led a successful prediction of the free parameters of the Standard Model.*[10]

*Now we may be at a new turning point, a radical change in what we accept as a
legitimate foundation for a physical theory. The current excitement is of course a consequence of the discovery of a vast number of solutions of string theory. Unless one can find a
reason to reject all but a few of the string theory 'vacua', we will have to accept that much
of what we had hoped to calculate are environmental parameters, like the distance of the
earth from the sun, whose values we will never be able to deduce from first principles.*

*We lose some, and win some. The larger the number of possible values of physical
parameters, the more string theory legitimates anthropic reasoning as a new basis for
physical theories: Any scientists who study nature must live in a part of the Multiverse
where physical parameters take values suitable for the appearance of life and its evolution
into scientists.*

*Some physicists have expressed a strong distaste for anthropic arguments. (I have heard
David Gross say 'I hate it') This is understandable. Theories based on anthropic calculation certainly represent a retreat from what we had hoped for: the calculation of all
fundamental parameters from first principles. It is too soon to give up on this hope, but
without loving it we may just have to resign ourselves to a retreat, just as Newton had to
give up Kepler's hope of a calculation of the relative sizes of planetary orbits from first
principles.* (Weinberg 2005)

---

[10] In the Standard Model equations about twenty quantities, 'parameters', appear which cannot be
calculated from first principles but must be extracted from experiments. Some of these are the
masses of the six leptons, the masses of the six, differently flavoured, quarks and the quantities
which determine the probability of the decays of heavy quarks and leptons into their lighter
cousins.

Particle Physics and Medical Physics

Figure 5 on p. 258 shows that we have come *half way* with the LHC (on a logarithmic scale) in the experimental study – based on the use of particle accelerators of larger and larger energy – of the phenomena which occurred between the Planck time and 380,000 years.

The vision of what could have happened before a millionth of a millionth of a second is hypothetical, but the overall picture is convincing because it is able to explain simply the state in which the cosmos is found 380,000 years later, when it was only a thousand times smaller than today.

I like to call this intellectual construction 'beautiful physics'; over the course of 70 years, tens of thousands of scientists – from accelerator and detector experts to experimental and theoretical physicists – have contributed to it.

Usually popular books give prominence to the theorists, who constructed the theories with which one can explain the data produced by accelerators or predict future observations. Less frequently they give the right weight to the experimenters, but almost never to accelerator technologists and detector experts; in the first three chapters of this book I tried to do them justice, recounting the lives and contributions of a few of them.

By way of compensation, I omitted to name many theoretical physicists, who have made essential contributions to the construction of the Standard Model. But this should not deceive the reader; beautiful physics is the result of the inventiveness and perseverance of engineers, experimental physicists and theorists who, over decades, have learned to work together and to understand each other, thanks also to the community life made possible by great accelerator laboratories like CERN, Fermilab, BNL, DESY and SLAC.

In parallel with the development of beautiful physics, many accelerator and detector experts have worked on medical applications of techniques and instruments, developed originally only to respond to the increasingly exacting requests posed by fundamental physics. I traced the history of applications of accelerators to cancer therapy in the final chapters, to offer a perspective on these fundamental spin-offs from particle physics.

The three particle accelerator stories are strictly intertwined and advances in one line have profited the other, and vice versa. This is a very good example of how, in the words of the Nobel Prize winner Burt Richter – whom I quoted in Chap. 3 – *"the road from science to technology is not the broad, straight highway that many would like to believe."* As he said in a talk he gave as President of the American Physical Society, the process is similar to the double helix of the DNA: the two strands of science and technology are inextricably linked and mutually essential.

My passion for these instruments came about from the dual, interlinked, use of accelerators, in basic research and in medical physics, and also gave rise to a maxim which I often quote: "physics is beautiful *and* useful".

References

Alvarez LW (1987) Alvarez: Adventures of a Physicist. Alfred P. Sloan Foundation Series

Amaldi E (1986) John Adams and his times. CERN 86–04, pp 1–15

Andrade EN da Costa (1964) Rutherford and the nature of the atom. Peter Smith Pub Inc, Gloucester MA, p 111

Barrow J (2009) Cosmic imagery: key images in the history of science. W. W Norton, New York

Budker GI (1993) Reflections and remembrances. In: BN Breizman, James Walter Van Dam (eds) Springer-Verlag, New York

Centre Européen de la Culture (1950) Résolution du 12 décembre, Meeting of the 12 Dec 1950. Archives of the Centre Européen de la Culture, Geneva

CERN Archives (1952) Telegram from some two dozen conference delegates to Isidor Rabi, 15 Feb

CERN PS (1960) Proton Synchrotron Machine Group, operation and development, Quarterly report no 1, Jan–Mar

Chalmarès G (1905) La radiographie aux armées en champagne. In *La Nature*, pp 99–102

Charpak G, Saudinos D (1993) La vie à fil tendu. Editions Odile Jacob, Paris

Childs H (1968) An American genius: the life of Ernest Orlando Lawrence. E.P. Dutton, New York

Cole FT (1994) O Camelot! A memoir of the MURA years. PRINT-94-0126

Cormack AM (1979) Early two-dimensional reconstruction and recent topics stemming from it. Nobel Lecture

Crease RP (1999) Making physics: a biography of Brookhaven National Laboratory, 1946–1972. University of Chicago Press, Chicago

Cucinotta FA, Durante M (2006) Cancer risk from exposure to galactic cosmic rays: implications for space exploration by human beings. Lancet Oncol 7(5):431–5

Curie M (1921) La radiologie et la guerre. Félix Alcan, Paris

Curie È (1938) Madame Curie. Gallimard, Paris

Dam JW (1896) The new marvel in photography. A visit to Professor Röntgen at his laboratory in Wurzburg. His own account of his great discovery. McClure's Magazine

David Jackson J, Panofsky WKH (1996) Edwin Mattison McMillan 1907–1991. A Biographical Memoir. National Academies Press, Washington, DC, Copyright

de Broglie L (1949) Message of Louis de Broglie read by Raoul Dautry, 9/12/49, p. 1–2. Archives of the Centre Européen de la culture, Geneva

del Regato JA (1993) Radiological oncologists: the unfolding of a medical specialty. Radiology Centennial Inc., Reston

Dowling T (2012) How to explain Higgs boson discovery. The Guardian – Shortcuts Blog

© Springer International Publishing Switzerland 2015

U. Amaldi, *Particle Accelerators: From Big Bang Physics to Hadron Therapy*,
DOI 10.1007/978-3-319-08870-9

Evans L (2013) Colliders unite in the linear collider collaboration, CERN Courier, 28 Mar

Feynman R (1990) QED: the strange theory of light and matter. Penguin, London

Fowler WB, Samios NP (1964) The omega-minus experiment. Scientific American 211. No 4, 36

Gamow G (1993) Mr Tompkins in paperback, illustrated by George Gamow and John Hookham. Cambridge University Press, Cambridge. Image reproduced with permission

Glashow S (1975) Quarks with color and flavor. Scientific American 233:38–50

Grubbe EH (1933) Priority in the therapeutic use of X-rays, Radiology XXI:156–162

Greene B (1999) The elegant universe: superstrings, hidden dimensions, and the quest for the ultimate theory. Copyright © 1999 by Brian Greene. Image used by permission of W.W. Norton & Company, Inc., New York

Haidt D (2005) The discovery of weak neutral currents. AAPPS Bull 15N1:47–50

Heilbron JL, Seidel RW (1990) A history of the Lawrence Berkeley Laboratory, vol I. University of California press, Berkeley

Heisenberg W (1969) Theory criticism and philosophy. In: From a life of physics. IAEA, Vienna, pp 31–46

Huff D (1954) A 10,000-ton cracker for invisible nuts. Popular Science, vol 164, Num. Bonnier Corporation

Johnsen K (1994) Odd Dahl 1898–1994, CERN Courier

Jongen Y (2013) European Patent Office films. www.epo.org/learning-events/european-inventor/finalists/2013/jongen.html

Kraft G (2013) History of the Heavy Ion Therapy at GSI. Posted on the encyclopedic website "The Health Risks of Extraterrestrial Environments" http://three.usra.edu/#section=main

Lawrence EO (1945) Transcription of a telephone conversation of 7 Nov 1945

Lawrence EO (1951) The evolution of the cyclotron. Nobel Lecture

Lee C (2007) Carbon ions pack a punch. CERN Courier

Maiani L, Bassoli R (2013) A caccia del bosone di Higgs. Mondadori Università, Milano

McMillan E (1973) Early Accelerators and Their Builders. IEEE Trans Nucl Sci NS – 20(3):8

Meer S Van der (1984) Stochastic cooling and the accumulation of antiprotons. Nobel lecture

Miller D (1993) A quasi-political explanation of the Higgs Boson; for Mr Waldegrave, UK Science Minister. Online reference: http://www.hep.ucl.ac.uk/~djm/higgsa.html

Mould RF (1993) A century of X-rays and radioactivity in medicine. Institute of Physics Publishing, London

Nishijima K (1965) Weak interactions at high energy. In: Yuan LCL (ed) Nature of matter, BNL 888. Brookhaven National Laboratory, Upton, pp 102–122

Tapscott E (1998) Nuclear medicine pioneer, Hal O. Anger, 1920–2005. Republished in: J Nucl Med Technol 2005, 33:251

Oppenheimer JR (1956) Physics tonight. Phys Today 9:10–13

Panofsky WKH (1997) Evolution of particle accelerators and colliders, in Beam Line - SLAC, Stanford, Spring 1997, p. 36–44

Powell CF (1972a) Selected papers of Cecil Frank Powell, Nobel laureate. In: Burhop EHS, Lock WO, Menon MGK (eds) North-Holland Publ. Co., Amsterdam/London, p 26

Powell CF (1972b) Selected papers of Cecil Frank Powell, Nobel laureate. In: Burhop EHS, Lock WO, Menon MGK (eds) North-Holland Publ. Co., Amsterdam/London, pp 343–344

Ramsey NF (1966) Early history of Associated Universities and Brookhaven National Laboratory. In: Brookhaven lecture series no 55. Brookhaven National Laboratory, Upton, pp 1–16

Richter B (2002) The role of science in our society. SLAC-PUB-9284

Röntgen W (1895) On a new kind of rays. In: Proceedings of the Würzburg Physical-Medical Society, 28 Dec 1895

Rovelli C (2014) La realtà non è come ci appare. Raffaello Cortina Editore, Milano

Rubbia C (1991) Edoardo Amaldi. 5 September 1908–5 December 1989. Biographical Memoirs of Fellows of the Royal Society 37:2–31

Russell LK (1896) Line on an X-Ray portrait of a lady, from Life Magazine

Salam A (1979) Gauge unification of fundamental forces. Nobel lecture

Schopper H (2009) LEP. The lord of the collider rings at CERN 1980–2000. Springer, Berlin/Heidelberg

Schutt RP (1971) Science physics and particle physics. In: Yuan LCL (ed) Elementary particles, science, technology and society. Academic Press Inc., New York/London, pp 1–48

Schwartz M (1962) Discovery story: one researcher's personal account. In: Maglic B (ed) Adventures in experimental physics. Volue Alpha – World Science Commun., Princeton, 1972, pp 82–85

Silverman A (2000) The magician: Robert Rathbun Wilson 1914–2000. CERN Courier

Smolin L (2007) The trouble with physics: the rise of string theory, the fall of a science, and what comes next. Mariner Books – HMH Book Clubs

Suits CG (1982) William David Coolidge. A biographical Memoir. National Academy of Sciences, Washington, DC

Tobias C, Tobias I (1997) People and particles. San Francisco Press, San Francisco

Touschek B (1977) Centro Linceo Interdisciplinare. Acc. Naz. Lincei, Roma, No 33

Trombley W (1959) Triumph in space for a 'Crazy Greek. In: Life, March 30, pp 31–34

UNESCO (1950) Records of the general conference of UNESCO, fifth session, Florence

Veltman MJ (1986) The Higgs Boson. Scientific American 255:76–84

Waloschek P (1994) The Infancy of Particle Accelerators: Life and Work of Rolf Wideröe. Friedrich Vieweg & Sohn Verlag, Brunswick/Wiesbaden

Weinberg S (1977) The forces of nature. American Scientist 65:171–176

Weinberg S (2005) Living in the multiverse. Talk at the symposium *Expectations of a Final Theory* at Trinity College, Cambridge, 2 Sept 2005. ArXiv:hep-th/0511037v1

Weisskopf VF (1977) The development of the concept of an elementary particle. In: Karimaki V (ed) Proceedings of the symposium on the foundation of modern physics, B1 No 14. Pub. Univ. of Johensuu, Loma-Koli, p 201

Wells PNT (2005) Sir Godfrey Newbold Hounsfield, Biographical Memoirs of Fellows of the Royal Society. Vol. 51, doi: 10.1098/rsbm.2005.0014, 1 Dec 2005

Wiener C, Hart H (eds) (1972) Exploring the history of nuclear physics. American Institute of Physics, New York, p 25

Wilczek F (2012) Who's on first? Relativity, time, and quantum theory. NOVA'S physics blog: the nature of reality. Public Broadcasting Service, Arlinton (VA-USA)

Wilczek F (2013) Trouble with physics: time to follow beauty? New Sci 217(2906):46

Wilson RR (1946) Radiological use of fast protons. Radiology 47(5):487–491

Wilson RR (1969) Congressional Testimony – 17 April 1969, AEC Authorizing Legislation Fiscal Year 1970: Hearings before the Joint Committee on Atomic Energy. 91st Congress of the U.S.A., 1st session on General, Physical Research Program, Space Nuclear Program, and Plowshare, 17–18 April 1969, part 1. Washington, DC: U.S. Government Printing Office, pp 112–118

Wilson RR (1997) Foreword to Advances in Hadron therapy. In: Amaldi U, Larsson B, Lemoigne Y (eds), Elsevier, Amsterdam. ix Wilson, 1996

Wilson R (2004) A brief history of the Harvard University cyclotrons. Harvard University Press, Cambridge, MA

Witten E (2004) When symmetry breaks down. Nature 429, pp 507–508

Wolf G (2001) The glorious days of physics: a Tribute to Björn H. Wiik and his physics. In: Antonino Zichichi (ed) Basics and highlights in fundamental physics. Proceedings of the international school of subnuclear physics, The Subnuclear Series – vol 37. World Scientific, Singapore

About the Author

Ugo Amaldi has worked at CERN for thirty-five years, participating in and leading numerous experiments in particle physics and giving, in particular, significant contributions to the study of weak interactions and the unification of the fundamental forces. For thirteen years he has been spokesperson of DELPHI - one of CERN's international collaborations at the accelerator LEP, predecessor of LHC.

Already in the 1960s at the Italian Health Institute (ISS) in Rome he developed an interest in medical physics. In 1992 he came back to this subject with the creation of the TERA Foundation, which promotes research in oncological hadron therapy. In 2011, as a result of Amaldi's initiative, the Italian National Hadrontherapy Center for Cancer Treatment CNAO was opened in Pavia.

In the last thirty years more than one third of the Italian high-school pupils have studied physics on his textbooks.

© Springer International Publishing Switzerland 2015   271
U. Amaldi, *Particle Accelerators: From Big Bang Physics to Hadron Therapy*,
DOI 10.1007/978-3-319-08870-9

Author Index

© Springer International Publishing Switzerland 2015
U. Amaldi, *Particle Accelerators: From Big Bang Physics to Hadron Therapy*,
DOI 10.1007/978-3-319-08870-9

Subject Index

© Springer International Publishing Switzerland 2015

277

U. Amaldi, *Particle Accelerators: From Big Bang Physics to Hadron Therapy*,
DOI 10.1007/978-3-319-08870-9

Printed by Printforce, the Netherlands